Transforming Europe in the Images of the World, 1110–1500

AF787634

Transforming Europe in the Images of the World, 1110–1500

Fuzzy Geographies

Natalia I. Petrovskaia

Routledge
Taylor & Francis Group
LONDON AND NEW YORK

The publication of this book is made possible by a grant from the Dutch Research Council.

First published in 2025 by Amsterdam University Press Ltd.

Published 2025 by Routledge
4 Park Square, Milton Park, Abingdon, Oxon OX14 4RN
605 Third Avenue, New York, NY 10158

Routledge is an imprint of the Taylor & Francis Group, an informa business

© N.I. Petrovskaia / Taylor & Francis Group, 2025

The Open Access version of this book, available at www.taylorfrancis.com, has been made available under a Creative Commons Attribution-NonCommercial-NoDerivatives (CC-BY-NC-ND) 4.0 International license.

Trademark notice: Product or corporate names may be trademarks or registered trademarks, and are used only for identification and explanation without intent to infringe.

ISBN: 9789048563166 (hbk)
ISBN: 9781041189770 (pbk)
ISBN: 9781003708537 (ebk)
NUR 684

Cover image: Beginning of the Europe section of *Imago mundi* Staats- und Stadtbibliothek Augsburg, 2 Cod 108, f. 65v; reproduced by kind permission of Staats- und Stadtbibliothek Augsburg

Cover design: Coördesign, Leiden

DOI: 10.5117/9789048563166

Every effort has been made to obtain permission to use all copyrighted illustrations reproduced in this book. nonetheless, whosoever believes to have rights to this material is advised to contact the publisher.

For Product Safety Concerns and Information please contact our EU representative:
GPSR@taylorandfrancis.com
Taylor & Francis Verlag GmbH, Kaufingerstraße 24, 80331 München, Germany

Table of Contents

List of Figures and Tables 7

Acknowledgments 9

List of Abbreviations 11

Introduction: *Starting Out.* 'Europes', Hippogriffs, and Mathematics 13
Europe: What's in a Name? 18
Of Space-Time, Borders, and Mathematics 24
A Roadmap for this Book 32

Part I An Introduction to the *Imago Mundi* tradition

1. ***Understanding the World.* An Overview of the *Imago mundi*** 37
Medieval 'Encyclopedias' 37
About the Author and Concerning Context 41
The Contents of the *Imago Mundi* 45

2. ***Translating Knowledge.* An Introduction to the *Imago Mundi* Family** 53
Vernacular Translations of *Imago mundi* 62
Adaptations of *Imago mundi* I and Insertion of Translated Extracts in Encyclopedic Contexts 67
Adaptations and Translated Extracts Drawn into Other Genres 74
Adaptation into Latin 77

Part II Modes of Reading Geography

3. ***Time.* Authority and Archaism** 83
Authority in Fragments 85
The Timelessness of Archaism 87
Overlapping Maps: The Cases of *paludes Meotides*, Spain, and Rome 94

4. *Space.* **Geographical Regions as Fuzzy Sets** 105
Divide or Describe? *Dirimere, deviser* 106
Regions as Fuzzy Sets 109
Overlap between Fuzzy Sets in the *Image du monde* 111
The Mediterranean as Fuzzy Set 121

5. *Movement.* **The Hodoeporical Descriptive Technique** 129
Reader, Listener, or Viewer? 130
Going through the Text 135
Significant Itinerary Variations in the Latin Text 144
Local Interest 151

Conclusion: ***Looking Back and Looking Ahead*** 155
Looking Back 158
Looking Ahead 166

Appendices 173
Appendix 1: List of *Imago Mundi* chapters 173
Appendix 2: Texts of the *Imago Mundi* Tradition 176

Bibliography 177
Manuscripts 177
Primary Sources 177
Secondary Sources 180
Electronic Resources 201

Index 203

List of Figures and Tables

Figure 1	Timeline of *Imago mundi* adaptations	59
Figure 2	Categories of *Imago mundi* adaptations visualised in an opacity-varying diagram	62
Figure 3	The chapter on Germany in Staatsbibliothek Preussischer Kulturbesitz Germ. Fol. 595, f. 85v detail	94
Figure 4	*Imago mundi* locations mentioned in Chapter 3	97
Figure 5	Narrative Progression of *Imago mundi* mapped onto a T-O schema	138
Figure 6	Narrative Progression of *Image du monde* mapped onto a T-O schema	139
Figure 7	The direction viewed from the perspective of reader inside the map	140
Figure 8	Map of Italy showing the position of Sections (1)–(4) of Imago mundi I.26	148
Figure 9	Map of Italy showing the order of description of its regions in the *Trésor*	152
Figure 10	beginning of the Europe section of the *Imago mundi* in Augsburg, Staats- und Stadtbibliothek Augsburg 2 Cod 108, f. 65v. Reproduced with kind permission of the Staats- und Stadtbibliothek Augsburg	169
Table 1	The order of Chapter 27 sections in the *Imago mundi*	145

Acknowledgments

The research presented in this monograph was made possible by a research grant from the Dutch Research Council (NWO) under the VENI talent scheme, for the three-year project 'Defining "Europe" in Medieval European Geographical Traditions'. The Open Access publication of this book was made possible by a further generous grant by the NWO Open Access Books programme. I would like to thank NWO also for granting an extension of the project for the writing of this book, and Utrecht University, the Department of Languages, Literatue and Communication, the Institute for Cultural Enquiry, and the Center for Medieval Studies for supporting this project. Kiki Calis was research assistant on the project and collected the data for the project's manuscript database. In the course of the project, invaluable guidance, help, support on all aspects of practicalities, logistics, finances, administration, and scholarship was provided without fail by so many colleagues that it would be impossible to list all names without doubling the wordcount of the volume.

For the book itself also it is impossible to name everyone to whom my gratitude extends. The generosity of colleagues and friends with their time and in sharing their work has been invaluable both before and also during the crisis that started in spring 2020, when access to literature was made more difficult by library closures. I am grateful to Sarah Centili for making her doctoral thesis available to me, and to Maria Teresa Rachetta for her article on Brunetto Latini. To Sif Ríkharðsdóttir I am grateful for sharing a pre-publication version of her article on the Norse Charlemagne material. I also wish to thank Robert Hoyland for helping alleviate the effects of my embarrassing illiteracy in Hebrew. Many thanks also to the reviewers for their insightful comments on the final draft of the book, transmitted to me via Amsterdam University Press. Any residual errors or inaccuracies are, of course, my own.

I would also like to thank Shannon Cunningham, Victoria Blud, Julie Benschop-Plokker, Chantal Nicolaes and their colleagues at Amsterdam University Press who have made the process towards publication of this book such a smooth and enjoyable one.

The support, patience, and generosity of many colleagues, teachers and friends in sharing their time, their research, their good humour, their friendship, made this book possible.

Thank you.

List of Abbreviations

BL	British Library
BnF	Bibliothèque nationale de France
CORDE	*Corpus Diacrónico del Español* http://corpus.rae.es/cordenet.html
DMF	*Dictionnaire du Moyen Français (1330–1500)* http://www.atilf.fr/dmf/
PL	Patrologiae Cursus Completus, Series Latina

Introduction: *Starting Out*. 'Europes', Hippogriffs, and Mathematics

Abstract: The introduction provides an overview of the objectives, method, and structure of the book. The first part explains the focus on Europe within the discussion of the *Imago mundi* and the geographical tradition constituted by its translations and adaptations. The second section introduces two theoretical concepts which will play a central role in the following discussion: the chronotope and fuzzy sets. These are subsequently used in the book to explain the texts' treatment of regions and borders. The introduction concludes with a roadmap for the book.

Keywords: *Imago mundi*; Europe; medieval geography; chronotope; space-time; fuzzy sets

The majestic polyphony of European cultural references that is Ludovico Ariosto's *Orlando Furioso* features a flight over and beyond Europe on the back of a hippogriff.

> Ben che Ruggier sia d'animo constante,
> né cangiato abbia il solito colore,
> io non gli voglio creder che tremante
> non habbia dentro piú che foglia il core.
> Lasciato avea di gran spazio distante
> tutta l'Europa, et era uscito fuore
> per molto spazio il segno che prescritto
> avea già a'naviganti Ercole invitto.
>
> Quello ippogrifo, grande e strano augello,
> lo porta via con tal prestezza d'ale...[1]

1 Ludovico Ariosto, *Orlando Furioso*, canto 6, stanza 17; Ludovico Ariosto, *Orlando Furioso: A Selection*, ed. by Pamela Waley (Manchester: Manchester University Press, 1975), p. 55.

Petrovskaia, N.I. *Transforming Europe in the Images of the World, 1110–1500. Fuzzy Geographies.* Taylor & Francis Group, 2025
DOI 10.5117/9789048563166_INTRO

> Courageous man that he was, Ruggiero's face retained its normal hue; but I do believe that his heart within him was trembling like a leaf. He had left the European mainland far behind him, and had passed way out beyond the bounds which matchless Hercules had set for mariners. The great and wondrous bird, the hippogryph, bore him away...[2]

This passage is both seemingly unproblematic and in reality impossible, and not only because of the hippogriff.[3] The latter is not the only imaginary entity involved. The other, somewhat unexpectedly perhaps, is 'Europe': for the first challenge in any discussion of Europe as 'geographical unit' or 'space' is the geographer's valid objection that there is no such thing.[4] Though frequently attached by convention in everyday discourse to the term 'continent', Europe is not a continent (if we can speak of continents at all), as Martin W. Lewis and Karen E. Wigen point out.[5] It does not coincide, for instance, with a single tectonic plate that would arguably provide a 'natural' boundary.[6] The term does not correspond, if one takes the extreme view, to a geographic object. The problem is epistemological, and also fundamentally a paradox: a term conventionally and habitually used, and for which people seem to agree, more or less, on a meaning, that does not appear to have a corresponding physical referent. This reflects Robert L. Fowler's more general observation that, 'Even if the maps we create only make sense to us – even if our north is in absolute terms really south, or even if there are no absolute terms at all – we are compelled to orient ourselves somehow,

2 Ludovico Ariosto, *Orlando Furioso*, trans. by Guido Waldman, Oxford World Classics (Oxford: Oxford University Press, 1974), p. 52.

3 Ruggiero's travels later take him full circle around the world, returning to Europe by flying westwards again. For a discussion of Ariosto's geography, see Federico Italiano, *Translation and Geography* (Abingdon: Routledge, 2016), pp. 32–50; the discussion of Ruggiero's hippogriff flight is at pp. 38–43. Italiano's discussion begins with a question similar to the one posed here, on the relationship between the text and the geographical 'reality' as perceived by the author. See also Jo Ann Cavallo, *The World Beyond Europe in the Romance Epics of Boiardo and Ariosto* (Toronto, Buffalo, and London: University of Toronto Press, 2013).

4 The point is made so frequently as to render a full list of references impossible. See, for example, Gerard Delanty, *Formations of European Modernity: A Historical and Political Sociology of Europe* (London: Palgrave Macmillan, 2019), p. 16; Michael Heffernan, *The European Geographical Imagination* (Stuttgart: Franze Steiner, 2006), pp. 12, 13. Cf. the introduction to Arnaud Brennetot and Muriel Rosemberg, 'Géographie de l'Europe et géographie de la construction européenne', *L'Espace Politique*, 19 (2013), http://journals.openedition.org/espacepolitique/2613 (accessed 25 March 2020).

5 Martin W. Lewis and Karen E. Wigen, *The Myth of Continents: A Critique of Metageography* (Berkeley: University of California Press, 1997), pp. 17, 21, 29–30.

6 Lewis and Wigen, *The Myth of Continents*, p. 34.

or we simply cannot function'.[7] The cultural construction of a geographical system thus lies at the very basis of any discourse as the foundation of our capacity to place ourselves in the world. This tension between, on the one hand, geographical entities (and their depictions) as cultural constructs, and on the other, geographical description as a model of the world onto which such cultural constructs are mapped, is at the heart of *Transforming Europe*.[8]

This book has two objectives. The first, addressed in Part I, is to explore the medieval European geographical system by bringing together for the first time in a single discussion various branches of an extensive textual geographical tradition deriving from a single source – the Latin encyclopedia *Imago mundi*, composed in the twelfth century by Honorius Augustodunensis (fl. *c.* 1090–*c.* 1140).[9]

This text, written in fairly straightforward Latin, is an accessible and structured compendium of established knowledge, and, in Scott D. Westrem's words, 'perhaps the most generally known book with an extended, if rudimentary, discussion of geography'.[10] It was translated (we will come back to

7 Robert L. Fowler, 'Encyclopaedias: Definitions and Theoretical Problems', in *Pre-Modern Encyclopaedic Texts Proceedings of the Second COMERS Congress, Groningen, 1–4 July 1996*, ed. by Peter Binkley (Leiden: Brill, 1997), pp. 3–29 (p. 7).

8 The tension is fundamental to the study of geography; see J. B. Harley, 'Maps, Knowledge, and Power', in *The Iconography of Landscape: Essays on the Symbolic Representation, Design and Use of Past Environments*, ed. by Denis Cosgrove and Stephen Daniels (Cambridge: Cambridge University Press, 1988), pp. 277–312; J. B. Harley, 'The Map and the Development of the History of Cartography', in *History of Cartography I: Cartography in Prehistoric, Ancient, and Medieval Europe and the Mediterranean*, ed. by J. B. Harley and David Woodward (Chicago and London: University of Chicago Press, 1987), pp. 1–42 (p. 3).

9 For editions, see *De imagine mundi libri tres*, ed. by Jean-Paul Migne, Patrologiae Latinae Cursus Completus (PL) 172 (Paris, 1895) and *Imago Mundi*, ed. by Valerie I. J. Flint, *Archives d'histoire doctrinale et littéraire du moyen Âge*, 49 (1982). It should be noted that this edition differs in its chapter divisions from the earlier ones. A translation of the text is available in Nicholas Ryan Foster, 'The *Imago mundi* of Honorius Augustodunensis' (unpublished MA thesis, Portland State University Department of History, 2008), online: DOI: 10.15760/etd.5974 (accessed 11 April 2024). Throughout the following discussion I use my own translation of the *Imago mundi*, since in some instances precise nuances in the wording of the Latin text, and in some instances, comparison with vernacular adaptations, required a separate translation. For a recent brief overview and bibliography relating to Honorius and his *Imago mundi*, see Michael W. Twomey, 'Honorius Augustodunensis (c. 1080–c.1140)', in *Routledge Resources Online: Medieval Studies*, DOI: 10.4324/9780415791182-RMEO96-1 (accessed 24 February 2023); see also *ARLIMA: Archives de littérature du moyen âge*, https://www.arlima.net/eh/honorius_augustodunensis (accessed 11 April 2024) and, for the *Imago mundi* in particular, Bayerische Akademie der Wissenschaften database *Repertorium: 'Geschichtsquellen des deutschen Mittelalter'*, https://www.geschichtsquellen.de/repOpus_02846.html (accessed 11 April 2024).

10 Scott D. Westrem, *Broader Horizons: A Study of Johannes Witte de Hese's Itinerarius and Medieval Travel Narratives*, Medieval Academy Books 105 (Cambridge, MA: Medieval Academy,

this term and the variety of concepts it represents) into multiple European vernaculars and incorporated in the form of extracts and quotations into both scientific treatises and literary works produced in different parts of Europe in various languages throughout the medieval period, becoming one of the most popular and influential medieval European encyclopedic texts.[11] Despite this enormous popularity, the *Imago mundi* is more often the subject of footnotes in contemporary scholarship than of separate studies (if it is mentioned at all). However, as we shall see in this book, this text's monumental importance deserves more attention. Thus, the opening chapters of this book constitute the first overview of the enormous and influential tradition generated by this text. The extended family of texts that translate, adapt, and use sections of *Imago mundi* in a new context is sufficiently broad as to make a complete analysis a multi-volume endeavour. This book therefore makes the first step towards the systematic study of this tradition as a single, multilingual and multi-genre whole.[12] It is hoped that it will facilitate further discussion.

2001), p. 20 n. 43. For an overview of the contents and structure of this text, see Chapter 1 below, and for an overview of its translations and adaptations, Chapter 2. For comments on the accessability of Honorius's style, see Flint, ed., *Imago mundi*, p. 14, and more generally, Walter Andrew Hannam, 'The *Ineuitabile* of Honorius Augustodunensis: A Study in the Textures of early Twelfth-Century Augustinianisms' (unpublished PhD dissertation, Boston College, Graduate School of Arts and Sciences, 2013), pp. 62–65, 89, 91, 135–37, 142, 194–95. See also Michael W. Twomey, 'Medieval Encyclopedias', in R. E. Kaske, *Medieval Christian Literary Imagery: A Guide to Interpretation* (University of Toronto Press, 1988), pp. 182–215 (pp. 189–91).

11 See Figure 1, p. 59. According to Michael W. Twomey the *Imago mundi* is one of 'only a small handful of encyclopedias that existed in sufficient numbers in medieval Britain to have a chance of exercising some influence *and* surviving the *Suppression*' – the other three texts being the encyclopedias of Isidore, Bartholomaeus Anglicus (Bartholomew the Englishman), and Rabanus Maurus; Twomey, 'Inventing the Encyclopedia', in *Schooling and Society: The Ordering and Reordering of Knowledge in the Western Middle Ages*, ed. by Alasdair A. MacDonald and Michael W. Twomey, Groningen Studies in Cultural Exchange, 6 (Leuven, Paris, and Dudley, MA: Peeters, 2004), pp. 73–92 (p. 81), emphasis Twomey's. Jason Baxter uses *Imago mundi* as an example of the phenomenon he describes as 'an explosion of Latin encyclopedic investigations, which became the inspiration for a whole series of later Latin and vernacular treatises' in the twelfth century; Baxter, '"Videmus nunc per speculum": The Mysticism and Naturalism of the Twelfth-Century *imago mundi*', *Haskins Society Journal*, 28 (2017), 119–42 (p. 119).

12 For a brief prolegomenon to the study the plurivocal multilingual tradition of Honorius's other bestseller, the *Elucidarium*, see Gleb Schmidt, 'From Manual to Best-Seller: The History of Honorius Augustodunensis's *Elucidarium*', in *Books of Knowledge in Late Medieval Europe: Circulation and Reception of Popular Texts*, ed. by Pavlína Cermanová and Václav Žůrek, Utrecht Studies in Medieval Literacy, 52 (Turnhout: Brepols, 2021), pp. 137–63 (pp. 145–49). For the text of the *Elucidarium*, see Yves Lefèvre, L'Elucidarium *et les lucidaires: Contribution, par l'histoire d'un texte, à l'histoire des croyances religieuses en France au Moyen Âge*, Bibliothèque des Écoles françaises d'Athènes et de Rome 180 (Paris: E. de Brocard, 1954), pp. 343–497.

The second objective of the book, addressed in Part II, is linked to this issue of the tradition's popularity. Why was this encyclopedic text copied so much? And why was it translated into so many languages? Why were extracts from it added to so many other texts of such varied genres? I propose here a possible explanation for the phenomenal spread of the *Imago mundi* in a triad of characteristics: archaism, fuzziness (technical term!), and the flexible hodoeporical structure of the text. (I use the term 'hodoeporical' to designate an itinerary-like perambulating structure, where the reader is taken through the regions discussed according to a particular trajectory, going via adjacent regions.) Chapters 3, 4, and 5 in Part II of this book are dedicated to demonstrating and exploring each of these characteristics through the analysis of a selection of the text's vernacular adaptations.

Fuzziness is used throughout this book in its formal sense, as an application of fuzzy set theory developed by mathematician Lotfi A. Zadeh, which is particularly apt for use in the humanities and social sciences, where linear borders and strict binary categorisation have limited applicability.[13] I introduce fuzzy set theory and its implications for the analysis of the geographical material in the *Imago mundi* tradition in a separate section of this introduction.[14]

I explore these characteristics of the material by focusing on how the description of Europe (or, even perhaps, Europes, plural) is transformed in various vernacular texts of the tradition.[15] The size of the extended family of *Imago* mundi–derived texts renders it particularly suited to showing how the geographical description of Europe was 'shifting, divisible and flexible',

13 Lotfi A. Zadeh, 'Fuzzy Sets', *Information and Control*, 8 (1965), 338–53. For a discussion of the slow application of fuzzy set theory in the humanities, anchored in analysis of differing scientific cultures in the two fields, see Settimo Termini, 'On Some "Family Resemblances" of Fuzzy Set Theory and Human Sciences', in *Soft Computing in Humanities and Social Sciences*, ed. by R. Seising, and V. Sanz González, Studies in Fuzziness and Soft Computing 273 (Berlin and Heidelberg: Springer, 2012), pp. 39–54. One of Termini's observations is that both soft computing (the broad field to which fuzzy set theory belongs) and the humanities require rigour and precision – but not digit-based numerical precision, which is often part of the hard sciences (ibid., p. 44).

14 See below, pp. 30–32

15 The tentative suggestion of a plural is inspired by Lucien Febvre's description of the Mediterranean as *un complexe de mers* ('a complex of seas') in Fernand Braudel's classic and pioneering study, *La Méditerranée et le Monde Méditerranéen à l'Epoque de Philippe II* (Paris: A. Colin, 1949); Febvre, 'Un livre qui grandit: *La Médieranée et le monde méditeranéen a l'époque de Philippe II*', *Revue historique*, 203.2 (1950), 216–24 (p. 218). For discussions of the importance of Braudel's work, see A. H. Merrills, *History and Geography in Late Antiquity* (Cambridge: Cambridge University Press, 2005), pp. 16–21; and Hugh Trevor-Roper, 'Fernand Braudel the Annales, and the Mediterranean', *The Journal of Modern History*, 4 (1972), 468–79.

to use Eric Hobsbawm's terminology, within the encyclopedic geographic tradition of the late Middle Ages, in the period of *c.* 1100 to *c.* 1500.[16]

In the following sections of this introduction, I first specify the focus of the enquiry – Europe – then introduce the theoretical framework based on fuzzy logic, and finally provide a roadmap for the book. Because the *Imago mundi* and the related family of texts are being examined together for the first time, the overview of this material is presented in dedicated chapters (Chapters 1 and 2), rather than in the introduction.

Europe: What's in a Name?

While the text quoted in the opening of this book, *Orlando Furioso* – almost surprisingly, given the wide range of the *Imago mundi* tradition – is not related to this family of texts in a direct way, the passage I quote from it vividly illustrates the implications and range of this enquiry.[17] If we can understand what Ariosto means when he has Ruggiero leave the space of Europe flying away on his winged steed, then 'Europe' in this passage must mean something geographical for author and audience, something spatial with a defined size and perhaps even boundaries one can cross.[18]

A way to cut this Gordian knot – a procedure needed in order to discuss geographical definitions of 'Europe' and permit the present book to be written – is offered by Gabrielle Spiegel in her discussion of poststructuralism's implications for literary and historical analysis: 'texts both mirror *and* generate social realities, are constituted by *and* constitute the social and discoursive formations which they may sustain, resist, contest, or seek to transform, depending on the case at hand'.[19] Thus, Europe as a geographic designation is very real because it is also a cultural (and in the twentieth and twenty-first centuries, also a political and a social) construct.[20]

Two particular challenges therefore face both the writer and her readers in a project such as this. The first challenge is that recent research has

16 Eric Hobsbawm, 'The Curious History of Europe', in *On History* (London: Weidenfield & Nicholson, 1997), pp. 217–27 (p. 219).

17 An indirect connection does, however, exist; see Chapter 5, p. 141.

18 For a discussion of the passage in *Orlando Furioso*, see Roberto M. Dainotto, *Europe (In Theory)* (Durham and London: Durham University Press, 2007), p. 43.

19 Gabrielle Spiegel, 'History, Historicism, and the Social Logic of the Text in the Middle Ages', *Speculum*, 65 (1990), 59–86 (p. 77), Spiegel's emphasis.

20 For the idea that Europe is the product of 'European geographical imagination', see Heffernan, *The European Geographical Imagination*, p. 8.

focused on the study of Europe not as a physical space, but as a social space, produced, in Henri Lefebvre's terms, through action and interaction, social and political.[21] According to Jacques Le Goff, 'Europe began as a myth and a geographical concept', and studies of the (medieval) idea of Europe often prioritize – justifiably given its contemporary relevance – the former over the latter.[22] For instance, Michael Wintle's monumental and finely crafted *The Image of Europe* and Klaus Oschema's *Bilder von Europa* provide an excellent overview of historical (and in the latter case, specifically medieval)

21 It is worth noting here the focus of many recent studies on the question of European identity: e.g. Chiara Bottici and Benoît Challand, eds, *Imagining Europe* (Cambridge University Press, 2014). See also observations in J. G. A. Pocock, *The Discovery of Islands: Essays in British History* (Cambridge: Cambridge University Press, 2005), p. 49. In Heikki Mikkeli's *Europe as an Idea and an Identity* (New York: St. Martin's Press, 1998), for instance, the focus is on 'ideas about Europe' (p. vi) but not on the geographical conception, and indeed geography is not mentioned at all in the preface. Some examples are discussed further below, but the bibliography is vast. While the history of the cultural and political meanings of 'Europe' lies beyond the realm of the present study, additional bibliographical references might be found in Klaus Oschema, *Bilder von Europa im Mittelalter*, Mittelalter-Forschungen, 43 (Ostfildern: Thornbecke, 2013); and Olaf Asbach, *Europa: Vom Mythos zur Imagined Community? Zur historischen Semantik 'Europas' von der Antike bus ins 17. Jahrhundert*, Europa und Moderne, 1 (Munich: Wehrhahn, 2011). For a discussion of the interrelation between the geographical and cultural constructs, see also Hobsbawm, 'The Curious History of Europe', and Henri Lefebvre, *The Production of Space*, trans. by Donald Nicholson-Smith (Malden and Oxford: Blackwell, 1991), esp. pp. 53, 57, 73. In this reading, a 'space' can be different in different periods of time. The work was originally published in French as *La production d'espace* (Paris: Anthropos, 1974).

22 Le Goff, *L'Europe est-elle née au Moyen Age?* (Paris: Seuil, 2003), p. 8. The book has been published in English as *The Birth of Europe* (Malden: Blackwell, 2005). Heffernan's observation, made several decades ago, still largely holds, despite the proliferation of studies dedicated to the idea of 'Europe': he argued that 'How Europeans have creatively imagined themselves *geographically* is a surprisingly ignored theme'. Heffernan, 'The Changing Political Map: Geography, Geopolitics, and the Idea of Europe since 1500', in *An Historical Geography of Europe*, ed. by R. A. Butlin and R. A. Dodgshon (Oxford: Clarendon Press, 1998), pp. 140–80 (p. 142), emphasis Heffernan's. Gerard Delanty observes that the distinction between the geographical and cultural meanings often becomes invisible; Delanty, *Inventing Europe: Idea, Identity, Reality* (Basingstoke: Macmillan 1995), p. 4. For counterexamples (studies where a geographical definition is provided and the problem discussed), see e.g. Anthony Pagden, 'Europe: Conceptualising a Continent', in *The Idea of Europe: From Antiquity to the European Union*, ed. by Pagden (Cambridge: Cambridge University Press, 2002), pp. 33–54, though his focus is not specifically on the Middle Ages; see also Asbach, *Europa*, p. 29. For a recent bibliography of the subject see Klaus Oschema's monumental *Bilder von Europa im Mittelalter*. I am grateful to Christoph Mauntel for first bringing Oschema's monograph to my attention. Two further relevant volumes have appeared during the preparation of the present book: Oschema, *'Europe' in the Middle Ages* (Leeds: ARC Humanities Press, 2023) and Christoph Mauntel, *Die Erdteile in der Weltordnung des Mittelalters: Asien – Europa – Afrika* (Stuttgart: Anton Hiersemann Verlag, 2023). Thanks are due, once more, to Christoph Mauntel for alerting me to these two new publications.

uses of the term 'Europe'.[23] Oschema's great achievement is demonstrating that not only was the term 'Europe' used quite extensively in medieval discourse, but that contrary to the assertion in Denys Hay's classic work the concept of 'Europe' was not subservient to the concept of 'Christendom' and thereby devoid of political meaning.[24] Hay's main thesis is that the term 'Europe' in the Middle Ages only had a geographical sense, and that the real equivalent and precursor of the modern cultural conception of Europe was the concept of 'Christendom'.[25] The idea that the meaning of the term in

23 Of particular relevance to the present study is Wintle's discussion of geographical boundaries of Europe; see *The Image of Europe: Visualising Europe in Cartography and Iconography Throughout the Ages* (Cambridge: Cambridge University Press, 2009), at pp. 41–52, 86–98, 164–77. Oschema's study is more narrowly focused on the medieval period. However, the multitude of sources presented in the course of his analysis, particularly within the chronologically structured first part of his opus, makes it difficult to isolate individual types of use; see also Albrecht Classen's and Klaus Herbers's reviews of the book in *Mediaevistik*, 27 (2014), 245–46, and *Zeitschrift für Historische Forschung*, 43 (2016), 77–78, respectively.

24 A similar argument to Oschema's is found in Claire Weeda, 'Images of Ethnicity in Later Medieval Europe' (unpublished PhD dissertation, Universiteit van Amsterdam, 2012), p. 68 n. 97; see also her *Ethnicity in Medieval Europe, 950–1250: Medicine, Power and Religion* (York: York Medieval Press, 2021). For importance of Hay's work see, for instance, Ingrid Baumgärtner, 'Europa in der Kartographie des Mittelalters. Repräsentationenen – Grenzen – Paradigmen', in *Europa im Weltbild des Mittelalters. Kartographische Konzepte*, ed. by Ingrid Baumgärtner and Hartmut Kugler (Berlin: Akademie Verlag, 2008), pp. 9–28 (p. 11); Wintle, *The Image of Europe*, pp. 33, 156–58, 161; Michel Mollat du Jourdin, *Europe and the Sea* (Oxford: Blackwell, 1993), pp. 115–16; Delanty, *Inventing Europe*, at, for instance, pp. 10, 14, 16–17, 33–42; Pagden, 'Europe', pp. 74–75. Echoes can also be seen in discussions of ideas of 'Europe' for later periods. See, for instance, Ezra Talmor, 'Reflections on the Rise and Development of the Idea of Europe Europe', *History of European Ideas*, 1 (1980), 63–66. For a discussion of the extent of Hay's influence on more recent scholarship, see Oschema, *Bilder von Europa*, pp. 75–76. Hay's work has even led to some suggestions that the geographical concept of Europe as was not used in the medieval context such (or at least the denial of any importance of the concept); see, for instance, William Chester Jordan, '"Europe" in the Middle Ages', in *The Idea of Europe*, ed. by Pagden, pp. 72–90 (pp. 74–75), and the discussion of this view in Oschema, *Bilder von Europa*, pp. 19–23 and 60–76.

25 *Europe: The Emergence of an Idea* (Edinburgh: Edinburgh University Press, 1957), p. x; Denys Hay, *The Medieval Centuries*, 2nd edn (London: Methuen, 1964), p. 3. The term 'Christendom' had a political as well as a spatial signification; see Nora Berend, 'The Concept of Christendom: A Rhetoric of Integration of Disintegration?', in *Hybride Kulturen im mittelalterlichen Europa*, ed. by Michael Borgolte and Bernd Schneidmüller (Berlin: de Gruyter, 2010), pp. 51–62 (p. 55). For a criticism of this idea, see Oschema, *Bilder von Europa*, esp. pp. 31–32. A striking example of the distinction between 'Europe' the territory and 'Christendom' as cultural unity is presented in the famous crusading speech of Urban II; the William of Malmesbury version is found in *De Gestis Regum Anglorum*, ed. W. Stubbs, Rolls Series 90, 2 vols (London, Her Majesty's Stationary Office, Rolls Commission, 1887–89), II, pp. 394–95; translated in Hay, *Europe: The Emergence of an Idea*, p. 32. This is discussed in Chapter 3 below. An example of the explicit association between the two concepts, written by the Mallorcan cartographer Guillem Soler in the 1380s, is discussed

the Middle Ages was exclusively geographical has since been questioned, marking a dominant trend in the scholarly studies on the subject: the study of the history and development of the current cultural concept of 'Europe'.[26] This underlines the importance of the model–construct tension within the history of geographical ideas that I mentioned at the outset of this discussion, and since the tension in this case concerns our understanding of medieval geographical constructs, it brings us to the medieval side of our problem.

The polyvalence of the word results in a situation where 'Europe', to quote J. G. A. Pocock, 'can be defined in so many ways that one must always ask – but not always hope to be told – in what sense it is being used at the moment.'[27] As Henk Wesseling points out 'what we understand by the word Europe is not a geographical unity'.[28] The term can mean different things in different contexts, including a rhetorical topos, a legendary female figure, a king, or 'a geographical entity' which, as Caroline D. Eckhardt puts it, 'could be mapped in abstract terms as one of the three inhabited continents

by Sandra Sáenz-López Pérez, in 'El "otro" en la cartografía bajomedieval: Aportaciones desde la lectura de los mapas', in *Através do olhar do Outro: Reflexões acerca de sociedade europeia (séculos XII–XV)*, ed. by José Albuquerque Carreiras, Giulia Rossi Vairo, and Kristjan Toomaspoeg (Tomar: Instituto Politécnico de Tomar, 2018), pp. 135–56 (p. 142). Voltaire, in his *Essai sur les moeurs et l'esprit des nations*, also associated the medieval idea of Europe with Christendom; see the discussion in Wolfgang Schmale, 'Europe: Eighteenth-Century Definitions', in *Bordering Early Modern Europe*, ed. by Maria Baramova, Grigor Boykov, and Ivan Parvev (Wiesbaden: Harrazzowitz Verlag, 2015), pp. 79–93 (p. 85).

26 See in particular P. Gautier Dalché, 'Représentations géographiques de l'Europe – septentrionale, centrale et orientale – au Moyen Age', in *Europa im Weltbild des Mittelalters: kartographische Konzepte*, ed. by Ingrid Baumgärtner and H. Kugler, Orbis Mediaevalis: Vorstellungswelten des Mittelalters, 10 (Berlin: Akademie, 2008), pp. 63–79 (p. 63); William Wallace, 'Where Does Europe End? Dilemmas of Inclusion and Inclusion', in *Europe Unbound: Enlarging and Reshaping the Boundaries of the European Union*, ed. by Jan Zielonka (New York: Routledge, 2002); and Pagden, ed., *The Idea of Europe*. There have been so many works dedicated to this subject that a full bibliography is impracticable here. Examples include Delanty's *Inventing Europe*; Mikkeli, *Europe as an Idea*. Also relevant here is the question of what would have been understood as geography in the Middle Ages; Keith Lilley, 'Geography's Medieval History: A Neglected Enterprise?', *Dialogues in Human Geography*, 1 (2011), 147–62. For a detailed overview of previous studies, see also Oschema, *Bilder von Europa*, pp. 19–23 and 60–76.

27 Pocock, *The Discovery of Islands*, pp. 48–49. See also the description of differences between what was meant by 'Europe' in different time periods and a comment on the perceived stability of the concept, which he observes is often 'largely unquestioned' even when it is one the concepts 'purported to be the objects of enquiry', in Alexis Wick, *The Red Sea: In Search of Lost Space* (Oakland: University of California Press, 2016), pp. 62–63. See also, however, the discussion and references in Oschema, *Bilder von Europa*, at p. 77.

28 H. Wesseling, *A Cape of Asia. Essays on European History* (Leiden: Leiden University Press, 2011), p. 92.

of the spherical earth'.[29] *Transforming Europe* deals specifically with this abstract concept.

The second challenge is that we all already have a particular (and subjective) idea of Europe in mind, formed by our own time and cultural context.[30] Indeed, in contemporary discourse, the term 'Europe' is endowed with an almost unparalleled complexity. Our current geographical concept of 'Europe' (even if we leave aside for the moment the use of the term as shorthand for 'European Union' in contemporary discourse) is a modern construct, and it is difficult to shed its influence. The 'sevenfold continental system' invoked in a modern mind by the concept of Europe as continent only took its current form towards the second half of the twentieth century.[31] Furthermore, the very size, shape, and layout of the 'continent' (irrespective of its land boundaries, which pose a separate problem that we will return to in Chapter 4), so deeply imbedded in the modern imagination, have been shown to be a geocultural construct.[32]

The Mercator projection, but one method among many of representing the spherical Earth on a flat surface, expands the size of the northern hemisphere, making Europe appear much larger than it in fact is, in comparison to Australia and New Zealand, for instance, which appear much smaller than they are in three-dimensional reality.[33] Thus the complex of associations

29 As excellently summarised in Caroline D. Eckhardt, 'One Third of the Earth? Europe Seen and Unseen in the Middle English Chronicles of the Fourteenth Century', *Comparative Literature*, 58.4 'The Idea of Europe' (2006), 313–38 (p. 324).

30 Dipesh Chakrabarty, *Provincializing Europe: Postcolonial Thought and Historical Difference* (Princeton, NJ: Princeton University Press, 2000), pp. xii–xiv, 43.

31 Lewis and Wigen, *Myth of Continents*, p. 21.

32 In my use of 'geocultural' here I follow Romm, 'Continents, Climates, and Cultures: Greek Theories of Global Structure', in *Geography and Ethnography: Perceptions of the World in Pre-Modern Societies*, ed. by Kurt A. Raaflaub and Richard J. A. Talbert (Chichester: Wiley–Blackwell, 2010), pp. 215–35 (p. 216). See also Christopher Dawson, *Understanding Europe* (Washington, DC: Catholic University of America Press, 2008; first edition London: Sheed and Ward, 1952), pp. 39–40; Lewis and Wigen, *The Myth of Continents*, esp. pp. 16–17, 21–25; Peter J. Yearwood, 'Continents and Consequences: The History of a Concept', *Journal of Global History*, 9 (2014), 329–56; Pagden, 'Europe', pp. 36–37, 45–46.

33 The Mercator projection, its influence on contemporary cartography, and critique of its cultural impact have been the subject of much discussion. See, for instance, Wintle, *The Image of Europe*, pp. 255–56; Jordan Branch, *The Cartographic State: Maps, Territory, and the Origins of Sovereignty* (Cambridge: Cambridge University Press, 2013), pp. 57–58; Michael Wintle, 'The Early Modern Iconography of Europe: Visual Images and European Identity', in *Contesting Europe: Comparative Perspectives on Early Modern Discourses on Europe, 1400–1800*, ed. by Nicolas Detering, Clementina Marsico, and Isabella Walser-Bürgler (Leiden and Boston: Brill, 2020), pp. 54–76 (pp. 61–65). For an overview of the controversies and debates, see Monmonier, *Drawing the Line: Tales of Maps and Cartocontroversy* (New York: Henry Holt, 1995), esp. pp. 16–22, and

of the geographical term itself is vastly different in the present day from that of its medieval counterpart.[34] As will become apparent in the course of *Transforming Europe*, medieval and modern ideas of the geographical entity designated 'Europe' do not necessarily entirely coincide, although, through constraints of chronology and causality, medieval ideas will have perforce played a role in the genesis of modern ones.

As Ingrid Baumgärtner observes, the search for a set meaning for the concept of Europe in the Middle Ages is complicated by the very real possibility that every writer had their own individual idea of it.[35] Karl Leyser's 1992 article provides an important prolegomenon to the analysis of the 'geographically and cosmologically oriented understanding of Europe', founded on the tripartite worldview (Asia, Africa, and Europe), shared by the most important early medieval authorities – Augustine, Isidore, and Orosius – and passed on to later writers.[36] Leyser brings the narrative up to the eleventh century, and offers a tantalising glimpse into the importance of understanding just how Europe was defined as a geographical space by various medieval writers. It is therefore appropriate that the narrative in this book starts with the twelfth century, with a work that relies heavily on these previous authorities.

The choice to focus on the combination of the *Imago mundi* tradition and descriptions of Europe is thus justified by its nodal position in the development of medieval encyclopedic thought, which means its idea of Europe will be one of the more dominant ones. As I have already mentioned, this text was translated, together with fragments of the encyclopedic text itself, into different cultures, languages, texts and genres, and even continued to circulate in the Age of Discovery and the Age of Print. A selection of its

Monmonier, *Rhumb Lines and Map Wars: A Social History of the Mercator Projection* (Chicago and London: University of Chicago Press, 2010), esp. pp. 138–50.

34 This point was made by Patrick Geary in relation to peoples and nations in *The Myth of Nations* (Princeton, NJ: Princeton University Press, 2002), pp. 13, 37–38.

35 Baumgärtner, 'Europa in der Kartographie des Mittelalters', p. 10, with reference to Bernd Schneidmüller, 'Europa im Mittelalter: Vorstellungen und Forschungsaufgaben', in *Integration und Transformation in Europa: Beiträge aus dem Forschungsschwerpunkt 'Integration und Transformation in Europa (ITE)'*, ed. by Heinz-Dieter Wenzel, Forschungsforum: Berichte aus der Otto-Friedrich-Universität Bamberg, 9 (Bamberg: Universitäts-Verlag Bamberg, 1999), pp. 6–16 (p. 7); and Schneidmüller, 'Die mittelalterlichen Konstruktionen Europas: Konvergenz und Differenzierung', in *'Europäische Geschichte' als Historiographisches Problem*, ed. by Heinz Duchhardt and Andreas Kunz, Veröffentlichungen des Instituts für Europäische Geschichte, Abteilung Universalgeschichte Beiheft 42 (Mainz: Philipp von Zabern, 1997), pp. 5–24 (p. 12).

36 Leyser, 'Concepts of Europe in the Early Middle Ages', *Past and Present*, 137 (1992), 25–47 (pp. 26–27).

translations and adaptations form the core of this study. Their fidelity to the original ranges from the very faithful to a complete restructuring and incorporation into a new encyclopedic format, and thus even when the term 'translation' is used here, it is in its broadest sense.[37]

Of Space-Time, Borders, and Mathematics

In the article quoted towards the beginning of this Introduction, Gabrielle Spiegel cites Bakhtin's articulation of the analytical principles she proposes.[38] In the following discussion, recourse is made to Bakhtin again, for a concept of paramount usefulness, the *chronotope*, which he defined as 'the intrinsic connectedness of temporal and spatial relationships'.[39] This concept has a triple role to play in the following discussion.

In the first place, it is useful for articulating the interconnectedness of space-time in the medieval worldview. It is optimal for the discussion of medieval world maps (*mappae mundi*) which, to quote Alfred Hiatt, can be read 'chronologically, as well as chorographically, in terms of time as well as space'.[40] *Mappae mundi*, particularly those of the so-called T-O type, provide a visual representation corresponding to, and also complementing the type

37 See Chapter 2. See also Joëlle Ducos, 'Que traduire en français? Traductions uniques et traductions multiples', in *Translation and Authority: Authorities in Translation*, ed. by Pieter de Leemans and Michèle Goyens, The Medieval Translator, 16 (Turnhout: Brepols, 2017), pp. 39–52 (p. 40); and discussion in Natalia I. Petrovskaia, 'Translation and Transmission of Texts in Medieval Europe: Two Aspects of Translatio', in *Literature, Science and Religion: Textual Transmission and Translation in Medieval and Early Modern Europe*, ed. by Manel Bellmunt Serrano and Joan Mahiques Climent, Problemata Literaria 88 (Kassel: Edition Reichenberger, 2020), pp. 359–90 (p. 360).

38 Spiegel, 'History', p. 83.

39 M. M. Bakhtin, 'Forms of Time and of the Chronotope in the Novel: Notes toward a Historical Poetics', in *The Dialogic Imagination: Four Essays by M. M. Bakhtin*, ed. by Michael Holquist, trans. by Caryl Emerson and Michael Holquist, University of Texas Press Slavic Series, 1 (Austin: University of Texas Press, 1981), pp. 84–258 (p. 84). For a discussion of the medieval unity of space-time, see below, pp. 27–28. For the use of the concept in relation specifically to medieval geographical sources, see, for instance, Dale Kedwards, 'Geography', in *A Critical Companion to Old Norse Literary Genre*, ed. by Massimiliano Bampi, Carolyne Larrington, and Sif Rikharðsdóttir (Cambridge: D. S. Brewer, 2020), pp. 127–44 (p. 128 and n. 4), without reference to Bakhtin.

40 Hiatt, 'Maps of Empires Past', in *Post-Empire Imaginaries? Anglophone Literature, History, and the Demise of Empires*, ed. by Barbara Buchenau, Virginia Richter, and Marijke Denger (Leiden and Boston: Brill, 2015), pp. 3–23 (p. 14). See also comments on time and medieval maps in Robert Bartlett, 'Heartland and Border: The Mental and Physical Geography of Medieval Europe', in *Power and Identity in the Middle Ages: Essays in Memory of Rees Davies*, ed. by Huw Pryce and John Watts (Oxford: Oxford University Press, 2007), pp. 23–36 (pp. 32–33). Since there

of information that we find in geographical texts such as Book I of the *Imago mundi*. The name of the T-O maps reflects the tripartite structure: they are based on the simple schemas dividing the inhabited world – depicted by a circle – into three parts.[41] Thus, both these maps and our text describe the inhabited world as separated into three parts: Asia, Europe, and Africa, treated in that order in the text.[42]

The interrelation of the *mappae mundi* 'world maps', on the one hand, and geographical texts on the other, has been much discussed.[43] While the text of the *Imago mundi* is not illustrated, it is, in a number of manuscripts,

is a connection between medieval *mappae mundi* ('world maps') and the *Imago mundi* texts, Hiatt's observation is equally valid for both; Hiatt, 'Maps of Empires Past', p. 15.

41 The current type designation for these maps (T-O) derives from their description by Leonardo di Stagio Dati in the fifteenth century in his *La sfera*, 3.11; see Dati, *La sfera: Libri Quattro in ottava rima*, ed. by Enrico Narducci (Milan, 1865; reprinted Bologna, 1975); Enrico Narducci, ed., *La sfera: Libri Quattro in ottava rima* (Milan, 1865; reprinted Bologna, 1975). See also Roberto Almagià, *Planisferi, carte nautiche e affini dal secolo XIV al XVII esistenti nella Biblioteca Apostolica Vaticana*, Monumenta Cartographica Vaticana, 1 (Vatican: Biblioteca Apostolica Vaticana, 1944), p. 118; David Woodward, 'Medieval *Mappaemundi*', in *History of Cartography I: Cartography in Prehistoric, Ancient, and Medieval Europe and the Mediterranean*, ed. by J. B. Harley and David Woodward (Chicago and London: University of Chicago Press, 1987), pp. 286–370 (p. 301); and Wojciech Iwańczak, 'Borders and Borderlines in Medieval Cartography', in *Frontiers in the Middle Ages: Proceedings of the Third European Congress of Medieval Studies (Jyväskylä, 10–14 June 2003)*, ed. by O. Merisalo and P. Pahta, Textes et Etudes du Moyen Age, 35 (Louvain-la-neuve: Féderation Internationale des Instituts d'Études Médiévales, 2006), pp. 661–72 (p. 662 n. 2).

42 For discussion of the order, see Natalia I. Petrovskaia, 'La formulation tripartite du monde dans les encyclopédies et textes littéraires du Moyen Âge', in *La Formule au Moyen Âge IV / Formulas in Medieval Culture IV*, ed. by E. Louviot, C. Garcia, and S. Morrison, ARTeM Atelier de Recherche sur les Textes Médiévaux, 31 (Turnhout: Brepols, 2022), pp. 197–217.

43 See, for instance, Uwe Ruberg, '*Mappae mundi* des Mittelalters im Zusammenwirken von Text und Bild', in *Text und Bild: Aspekte des Zusammenwirkens zweier Künste in Mittelalter und früher Neuzeit*, ed. by Christel Meier and Uwe Ruberg (Wiesbaden: Ludwig Reichert, 1980), pp. 550–92; David Woodward, 'Medieval *Mappaemundi*', pp. 286–87; Patrick Gautier Dalché, 'De la glose à la contemplation: Place et function de la carte dans les manuscrits de haut Moyen Âge', in *Testo e immagine nell'alto medioevo: 15–21 aprile 1993*, Settimane di studio del Centro Italiano di Studi sull'alto Medioevo, 41 (Spoleto: Presso la sede del Centro, 1994), pp. 693–771; Patrick Gautier Dalché, 'Maps in Words: The Descriptive Logic of Medieval Geography from the Eighth to the Twelfth Century', in *The Hereford World Map: Medieval World Maps and their Context*, ed. by P. D. A. Harvey (London: BL, 2006), pp. 223–42; Margriet Hoogvliet, '*Mappae mundi* and Medieval Encyclopaedias: Image versus Text', in *Pre-Modern Encyclopaedic Texts*, ed. by Binkley, pp. 63–74; Silvère Menegaldo, 'Géographie et imaginaire insulaire au Moyen Âge, d'Isidore de Séville à Jean de Mandeville', *Les Lettres romanes*, 66 (2012), 37–86; Bettina Schöller, 'Transfer of Knowledge: *Mappae Mundi* Between Texts and Images', *Peregrinations: Journal of Medieval Art and Architecture*, 4 (2013), 42–55; Iwańczak, 'Borders and Borderlines in Medieval Cartography', p. 665.

accompanied by diagrams of the T-O *mappa mundi* type or wind schemas.[44] Like the synchronic representation of history in the *mappae mundi*, the geographical descriptions of Europe in the *Imago mundi* tradition cannot be divorced from the temporal references made within those descriptions. This is most striking where an extract of the *Imago mundi* text is incorporated into a narrative or historical text. The focus of the present book is therefore on the *chronotope* of Europe in the *Imago mundi* tradition.

The second and third important aspects of the *chronotope* for our purposes here are linked, and I discuss them together. Both derive from the specifics of Bakhtin's definition of the term, which is therefore worth quoting *in extenso*: 'We will give the name *chronotope* (literally "time space") to the intrinsic connectedness of temporal and spatial relationships that are artistically expressed in literature. This term [space-time] is employed in

44 At least seven manuscripts have T-O diagrams; Evelyn Edson, *Mapping Time and Space: How Medieval Mapmakers Viewed their World*, British Library Studies in Map History, 1 (London: BL, 1997), pp. 113, 182–83 n. 56. In Cambridge, Corpus Christi College MS 66 (s. XII), a detailed *mappa mundi* precedes the *Imago mundi*: see *Parker on the Web*, https://parker.stanford.edu/ (accessed 9 April 2024). It has links to the Hereford *mappa mundi*, connecting the *Imago mundi* to the cartographic tradition; Flint, ed., *Imago mundi*, p. 10; Michael W. Twomey, 'Honorius Augustodunensis', *Trade, Travel, and Exploration in the Middle Ages: An Encyclopedia*, ed. by John Block Friedman and Kristen Mossler Frigg, with Scott D. Westrem and Gregory G. Guzman (New York and London: Routlledge, 2000), pp. 259–61 (p. 260). In Exeter, Cathedral Library MS 3514 (s. XIII), p. 53, a *mappa mundi* follows the text; see Julia Crick, 'The Power and the Glory: Conquest and Cosmology in Edwardian Wales (Exeter, Cathedral Library, 3514', *Textual Cultures, Cultural Texts*, ed. by Orietta da Rold and Elaine Treharne (Cambridge: Brewer, 2010), pp. 21–42; and Natalia I. Petrovskaia, *Medieval Welsh Perceptions of the Orient* (Turnhout: Brepols, 2015), pp. 8, 11, and plate 1. An almost identical map is found in the Welsh-language Oxford, Jesus College MS 20 (s. XIV^{2}), f. 32v, which contains genealogies alongside narrative and religious texts; see Digital Bodleian https://digital.bodleian.ox.ac.uk/objects/93d67972-f107-40ee-a0a8-5327e827e812/ (accessed 27 February 2023); Diana Luft, Peter Wynn Thomas, and D. Mark Smith, eds, *Rhyddiaith Gymraeg 1300–1425* (2013) http://www.rhyddiaithganoloesol.caerdydd.ac.uk/cy/ms-home.php?ms=Jesus20 (accessed 27 February 2023). The French *Image du monde* contians T-O diagrams and it has been observed that its geographical section can be projected onto a *mappa mundi* with a slight adjustment of the T-shape; Chantal Connochie-Bourgne, 'Limites et diversités de l'Europe: Le parti pris par Gossouin de Metz (*Image du monde*, 1245)', in *De la Chrétienté à l'Europe: Actes du Colloque Orléans, mai 1993*, ed. by Bernard Ribémont (Orleans: Paradigme, 1995), pp. 49–62 (p. 54); cf. Chapter 4. The integration of illustrations in this text contrasts with the Latin tradition; Gossouin de Metz, *L'Image du monde, une encyclopédie du XIIIe siècle: edition critique et commentaire de la première version*, ed. by Chantal Connochie-Bourgne, unpublished PhD thesis, University of Paris, Sorbonne, Paris IV, 1999,p. 808, ll. 1845–46. Cf. references to accompanying illustrations in Isidore's *De natura rerum*, X.2, XI.1; Chet Van Duzer, 'A Neglected Type of Medieval Mappamundi and Its Re-imaging in the Mare Historiarum (BnF MS Lat. 4915, Fol. 26v)', *Viator*, 43 (2012), 277–301 (p. 281 n. 19).

mathematics, and was introduced as part of Einstein's Theory of Relativity'.[45] One important point that arises from this description is that the *chronotope* reflects the unexpected similarity between medieval and modern conceptualisations of space-time (an apparent paradox). Einstein's articulation of the theory is more physical than mathematical, but Bakhtin's description of space-time as first and foremost a mathematical concept is accurate. The idea of relative time was given mathematical form by Henri Poincaré around the same time as Einstein and by Hermann Minkowski a little earlier.[46]

There is a tendency in the history of physics to skip from Aristotle directly to Galileo and Newton as a continuum of scientific thought.[47] From the perspective of of mathematics and physics as modern disciplines, this historical account is accurate, as early modern theorists relied on Aristotle but not on medieval thought. However, from the perspective of historical study this results in a distortion, since the medieval period becomes a blip in the story ('dark' ages invisible in the narrative, much like 'dark' matter is in the universe). One wonders as a consequence whether a focus on philosophical conceptualisation rather than mathematical and physical description of space might not yield an alternative continuum of conceptual thought between the makers of the *mappae mundi*, on the one hand, and Minokwski, Poincaré, and Einstein on the other, wherein Galileo and Newton – whose theories are based on the idea of absolute time – would be the blip?

I make this observation here because it is an extension into the broader history of science (relating to space) of the comment made by J. B. Harley in relation to the history of cartography. Disucussing the prioritisation of scientific progress in accuracy in the history of mapmaking, Harley writes:

45 Bakhtin, 'Forms of Time', p. 84; emphasis and brackets (including square brackets) are Bakhtin's.

46 As Sir Roger Penrose points out, Einstein was not the first to propose the idea; Penrose, *The Road to Reality: A Complete Guide to the Laws of the Universe* (London: Vintage Books, 2005), pp. 383, 406; cf. Stephen Hawking, *A Brief History of Time: From the Big Bang to Black Holes* (London: Bantam, 2016), p. 23. Note that, given what we know of the conceptualisation of space-time on medieval maps, one might perhaps take issue with the notion that in the pre-modern period as a whole, 'Time was completely separate from and independent of space'; Hawking, *A Brief History of Time*, p. 21.

47 See, for instance, Penrose, *Road to Reality*, pp. 383–90. Compare Bülent Atalay, *Math and the Mona Lisa: The Art and Science of Leonardo da Vinci* (New York: HarperCollins and the Smithsonian Institution, 2006), p. 55. His statement 'Science [...] has progressed in fits and starts, its course sometimes entirely retrograde in direction' is followed by a reference to the 'reinvention' of science in the Renaissance and the beginning of a union between science on the one hand and technology on the other in the seventeenth century (ibid.). The implication appears to be that the Middle Ages represented a step backwards in terms of what is considered scientific progress.

> Taken alone, however, this aspect fails to provide a balanced view of the developments of maps in history. It assumes a linear historical progression and, moreover (somewhat anachronistically), assumes that accuracy of measurement and comprehensiveness were as important throughout the past as they have been in the modern period.[48]

The point I am making here differs somewhat from Harley's, because it is less of a critique and more of a health warning. A map is always an approximation, and a map made for the history of physics might justifiably skip some landmarks crucial for the history of geography.[49] I would suggest that the similarity between the four-dimensional space-time of the theory of relativity and the unified nature of medieval space-time which becomes visible by virtue of Bakhtin's *chronotope* concept might be one such landmark.[50] It can help us to understand the *Imago mundi* and the texts discussed in this book if we think of medieval space also as four-dimensional.

This brings me to my third point. Bakhtin's use of modern mathematical concepts as epistemic tools for analysis of literary phenomena provides inspiration for the approach taken here in my use of fuzzy sets as an analytical model. In order to introduce this model, I start by explaining the primary problem it seeks to address: the problem posed by the idea of (linear) boundaries between geographical regions.

The issue of geographical boundaries of Europe in the Middle Ages and their perception have been the focus of increasing attention recently.[51] This

48 Harley, 'The Map and the Development of the History of Cartography', p. 3. (My interpolation is in square brackets; the parentheses are Harley's.) Cf. also P. D. A. Harvey, 'Medieval Maps: An Introduction', in *History of Cartography I*, ed. by Harley and Woodward, pp. 283–85.

49 Omissions can be, but not always are deliberate; see the fundamental discussion in J. B. Harley, 'Silences and Secrecy: The Hidden Agenda of Cartography in Early Modern Europe', *Imago mundi*, 40 (1988), 57–76.

50 The unified reading of space-time in the medieval context is crucial to Georg Jostkleigrewe's explanation of the mysterious apparent shift in Europe's boundaries in the medieval French *Image du monde*; Jostkleigrewe, 'L'espace entre tradition et innovation: La géographie symbolique du monde et son adaptation par Gossouin de Metz', in *Construction de l'espace au Moyen Age: Pratiques et représentations. XXXVIIe Congrès de la SHMES. Mulhouse, 2–4 juin 2006* (Paris: Publications de la Sorbonne, 2007), pp. 369–78. I discuss this in detail in Chapter 4.

51 For instance, Baumgärtner and Kugler, eds, *Europa im Weltbild des Mittelalters*; Bernd Schneidmüller, 'Die mittelalterichen Destillationen Europas aus der Welt', in *Europa in der Welt des Mittelalters: Ein Colloquium für und mit Michael Borgolte*, ed. by Tillman Lohse and Benjamin Scheller (Berlin: De Gruyter, 2014), pp. 11–32, with reference to the *Imago mundi* on p. 14; Klaus Herbers, 'Europa und seine Grenzen im Mittelalter', in *Grenzräume und Grenzüberschreitungen im Vergleich: Der Osten und der Westen des mittelalterlichen Lateineuropa*, ed. by Klaus Herbers and Nikolas Jaspert, Europa im Mittelalter, 7 (Berlin: Akademie Verlag, 2007), pp. 21–41. Note

coincides with an increase in appreciation of the geographical information carried by both geographical texts and *mappae mundi*. Consequently, discussions of medieval boundaries, geographical and otherwise, have become progressively nuanced.[52] There is a greater awareness of the fact that medieval borders were not always clearly fixed or defined.[53] For example, as Giles Constable observes, as 'an intellectual construct existing territorially only as a western extension of the Asiatic land-mass', Europe is an illustrative 'example of the uncertainty of medieval frontiers'.[54] The connection between these concepts and the separation of 'self' and 'other' is not simple, as Constable points out.[55] Constable further urges us to 'question the paradigm of the Middle Ages as a period filled with frontiers and boundaries'.[56]

It has also been argued recently that the concept of the border as a political boundary or frontier in the linear sense, separating different cultures,

that Herbers departs from a discussion of geographical concepts and boundaries into discussion of political frontiers and an overview of historiography relating to the socio-politico-cultural concept of Europe.

52 For examples of nuanced approaches, see John Block Friedman, 'Cultural Conflicts in Medieval World Maps', in *Implicit Understandings: Observing, Reporting, and Reflecting on the Encounters between Europeans and Other Peoples in the Early Modern Era*, ed. by Stuart B. Schwartz (Cambridge: Cambridge University Press, 1994), pp. 64–95; Sharon Kinoshita, *Medieval Boundaries: Rethinking Difference in Old French Literature* (Philadelphia: University of Pennsylvania Press, 2006); Neil Cartlidge, ed., *Boundaries in Medieval Romance* (Woodbridge: Boydell & Brewer, 2008); David Abulafia and Nora Berend, eds, *Medieval Frontiers: Concepts and Practices* (New York and Abingdon: Routledge, 2016).

53 Giles Constable, 'Frontiers in the Middle Ages', in *Frontiers in the Middle Ages*, ed. by Merisalo and Pahta, pp. 3–28 (p. 9). However, as Constable remarks, clear demarcation through 'verbal mapping' would have been present in, for instance, charters (ibid.). Some medieval maps of Europe, such as that in Ralph Higden's *Polychronicon*, have been shown to emphasise internal over external boundaries; see Eckhardt, 'One Third of the Earth?': 'This impetus towards representational map-making gives less recognition to the continents; Europe's constituent elements may be visible, but the perimeters are unmarked, and continental identities as such are effaced by other structural patterns' (p. 324). Some, such as the map of the *Liber Floridus* (*c.* 1120) of Lambert of Saint-Omer, preserved in Ghent, University Library MS 92, f. 241v, seem to use natural boundaries (such as mountains and rivers) to separate out individual regions; see discussion in Camille Serchuk, 'Gaul Undivided: Cartography, Geography and Identity in France', in *Mapping Medieval Geographies: Geographical Encounters in the Latin West and Beyond*, ed. by Keith Lilley (Cambridge: Cambridge University Press, 2013), pp. 175–200 (pp. 192–94); the map is reproduced on p. 193.

54 Constable, 'Frontiers', p. 7.

55 Constable, 'Frontiers', p. 6. Since this text and its vernacular versions are occasionally accompanied by T-O maps, as mentioned above, the lack of a focus on boundaries within the textual version of the geographical description brings into question the assumption regarding the prominence of such boundaries in the pictorial representations also.

56 Constable, 'Frontiers', p. 6.

originated in thirteenth-century Spain, post-dating the origins of the T-O tradition and the *Imago mundi*.[57] The fact that, as we shall see, most of this text's vernacular adaptations date to the thirteenth century and later makes the issue of boundaries relevant to their discussion as an idea potentially emerging in that period. In this book, I will take this line of thought further, bringing into the field two mathematical concepts – 'set' and 'fuzziness' – and argue that the borders of Europe in *Imago mundi*–based texts are difficult to define because they are not there yet.[58] They are fuzzy areas occupying indeterminate space between described areas.[59]

'Fuzzy' is used here as a technical term and is key to understanding how geographical composition works in the *Imago mundi* tradition. Its advantage is that it allows for plurality of meaning. Fuzzy sets are useful for describing categories that do not have clean lines of separation, and areas without tidy boundaries. In *Transforming Europe* this applies not only to geographical descriptions of Europe but also to categories of *Imago mundi*–based texts (discussed in Chapter 2). Fuzzy sets are therefore a concept used heavily in this book to help us make sense of the material.

Sets, as defined by Paul R. Halmos, 'have *elements* or *members*. An element of a set may be a wolf, a grape, or a pigeon. It is important to know that a set itself may also be an element of some other set'.[60] It is useful to think of the geographical regions described in *Imago mundi* and its adaptations as sets which have components, rather than, as one might instinctively do – under the influence of later cartographic habits – as sharply delineated geometric areas. Indeed, in the transformative translation processes of 'Europe' mapped in this book, shifts in the set's membership make it is more useful to think of the region not in terms of the traditional kind of set (to

57 For more, see Nora Berend, 'Preface' to *Medieval Frontiers*, ed. by Abulafia and Berend, pp. x–xvi (p. xii). According to Berend, cartographically, linear frontiers start to emerge from the fourteenth century onwards; ibid., p. xiii. See also Constable, 'Frontiers'.

58 François de Medeiros observed this lack of boundaries in the *Image du monde* and Brunetto Latini's *Trésor*, both part of the *Imago mundi* family; *L'Occident et l'Afrique (XIIIe–XVe siècle): Images et représentations* (Paris: Éditions Karthala and Centre de Recherches Africaines, 1985), p. 70 n. 23. Cf. Connochie-Bourgne, 'Limites et diversités', p. 59 n. 6. Compare also Ralph W. Brauer on medieval Arabic geography, in *Boundaries and Frontiers in Medieval Muslim Geography* (Philadelphia: American Philosophical Society, 1995). Brauer's findings are summarised nicely in James A. Miller's review in the *International Journal of Middle East Studies*: Brauer 'finds that their world was not one of abrupt political lines, but, rather, focused on capital cities and centres of power, identity, and control' (pp. 439–40).

59 I draw here on Lefebvre's concept of space as a social and cultural production, where space needs to be experienced in order to be constructed; *The Production of Space*, pp. 26, 31, 34–35.

60 Halmos, *Naïve Set Theory* (Princeton, NJ: Princeton University Press, 1960), p. 1. Emphases Halmos's.

which things either belong or do not belong) but rather as a fuzzy set, for which membership is a question of gradation.[61] Fuzzy sets were developed in mathematics originally as a model for non-determinate data, that is, where the issue of whether entities belong to a set is not a binary decision (yes or no) but expressed in degrees.[62] Lotfi A. Zadeh, who first introduced the concept, defines the fuzzy set thus: 'a "class" with a continuum of grades of membership'.[63] Although it has as yet found no wide application in the humanities (though it is used in social sciences), fuzzy set theory offers a flexible epistemic tool which allows for rigorous analysis of seemingly disordered and non-uniform material.[64]

Because the fuzzy set is a concept created to help construct a framework that 'provides a natural way of dealing with problems in which the source of imprecision is the absence of sharply defined criteria of class membership rather than the presence of random variables', to quote Zadeh again, it is particularly useful to think of the areas of the world (e.g. Europe) in the *Imago mundi* texts as fuzzy sets, which might have intersections ('the intersection of *A* and *B* is the *largest* fuzzy set which is contained in both *A* and *B*', where A and B are fuzzy sets themselves).[65] This also provides a useful paradigm for phenomena that have been observed by previous scholars. As Heffernan points out, 'the medieval order was based on multiple loyalties and complex allegiances operating in an overlapping and essentially a-spatial fashion', with the result that any 'idea of compartmentalized political space was, therefore, somewhat alien to the medieval Christian worldview'.[66] Thus, the medieval map, visual or textual, cannot easily draw linear political

61 For an accessible introduction to the concepts of a traditional type of set, with its strict dichotomy of members and non-members, and fuzzy set, which allows for degrees of membership, see Witold Pedrycz and Fernando Gomide, *An Introduction to Fuzzy Sets: Analysis and Design* (Cambridge, MA: MIT Press, 1998), pp. 3–8.

62 The use of the concept of the 'fuzzy set' borrowed from mathematics in relation to medieval geography is inspired by David Abilafia's use of the term – without reference to mathematics – in relation to boundaries. Abilafia writes: 'Fuzzy boundaries may be less obvious in the twenty-first century; but they certainly are ever-present'; Abulafia, 'Introduction' to *Medieval Frontiers*, ed. by Abulafia and Berend, p. 17. The concept of the fuzzy set was developed by Zadeh and subsequently developed into a more general form of fuzzy logic by Joseph Giguen; for an introduction to fuzzy logic and a comparison with traditional, binary logic, see e.g. Merrie Bergman, *An Introduction to Many-Valued and Fuzzy Logic: Semantics, Algebras and Derivation Systems* (Cambridge: Cambridge University Press, 2008), esp. pp. 2, 9, 117, for definitions.

63 Zadeh, 'Fuzzy Sets', p. 339.

64 Termini, 'On Some "Family Resemblances"', pp. 42–43 for quotation and discussion of Lofti Zadeh's querying the situation in an interview.

65 Zadeh, 'Fuzzy Sets', p. 339; see p. 341 for intersections. Emphasis in the original text.

66 Heffernan, 'The Changing Political Map', p. 144.

boundaries, because each space might have different overlapping connections to surrounding spaces. We return to the notion of overlap in Chapter 4, where we discuss the fuzziness of the boundary between Europe and Africa, which shifts between the *Imago mundi* and its French adaptation.[67]

Fuzzy sets represent the most experimental aspect of this book. The concept is used to explain one aspect of the flexibility that accounts for the success and popularity of the geographical description provided by the *Imago mundi*. Another aspect of this flexibility is the timeless nature of the text's geography. We also will look at the changes made to the text by its adaptors, specifically those dictated by the cultural and political context of adaptation. In framing the narrative I draw on the time-honoured Dantean model used by some of the more distant relatives of the *Imago mundi*, such as Fazio degli Uberti's *Dittamondo*, and take Ariosto's Ruggiero as our guide on the journey through this textual tradition. We will also see in Chapter 5 that this idea of the text's audience figuratively journeying through the text is reflected in both the structure and the terminology of several of the texts in this family. Let us now turn to our itinerary.

A Roadmap for this Book

Ruggero's deceptively simple act of displacement in the passage quoted at the start of this Introduction, and the surprising interpretative difficulties it carries, will be our guide in this book. We will ask what *tutta l'Europa* ('all of Europe') meant to medieval readers, using a Latin geographical encyclopedia as our case study.

The five chapters of this book are organised into two parts of the book. Part I introduces the material. This is necessary because this is the first study to look at the multilingual tradition of the *Imago mundi* as a whole. Chapter 1 introduces the notion of a medieval 'encyclopedia', the *Imago mundi*, its origins, and its contents. In Chapter 2 we will examine the texts that transmitted this knowledge across a wide geographic, cultural, and linguistic range with an overview of the text's vernacular translations and adaptations.

Part II builds on this foundation to explore why it is that the encyclopedia introduced in Chapter 1 gave rise to the massive tradition outlined for the reader for the first time in Chapter 2. The three chapters in Part II thus focus on three reasons why I think the Latin text became so incredibly popular.

67 See pp. 111–21.

The three Chapters in this Part can be broadly labelled as Time, Space, and Movement. Chapter 3 is *Time*, and examines the marked archaism of the *Imago mundi*, expressed by its use of Roman provincial nomenclature for territories that had long since acquired different denominations. This archaism is inherited by most of the text's adaptations, though some update information regarding particular areas.[68] We then examine the construction of the geographical model in Chapter 4 (*Space*), where fuzzy sets are used to make sense of several problematic aspects of the tradition. We will explore how our tendency to describe geographical entities by delineating their borders is sometimes at odds with the fuzziness of medieval texts. In this, too, Ariosto's Ruggiero helps us. In the passage quoted at the beginning of this book, the English word 'bounds' with its implications of a border(line) translates the Italian *segno* ('sign'), which refers to the Pillars of Hercules, marking the end of explored territories. My point here is not to question the accuracy of the translation but to highlight the potential space between the text and translation that tantalisingly opens up possibilities for new interpretative frameworks, such as the one introduced in Chapter 4.

Chapter 5 explores how some of the vernacular translations and adaptations of the *Imago mundi* reorder the text to walk the reader through the world (almost literally). I call this image of walking the reader through the text the 'hodoeporical descriptive technique'. We will examine the routes by which the readers are taken through the various territories described in the different geographical texts. We will conclude with a glimpse into the tradition's post-medieval future, returning to the early modern world of Ariosto and his Ruggiero.

Let us now turn to Chapter 1 and our starting point, the *Imago mundi* and the idea of a medieval 'encyclopedia'.

68 Flint, ed., *Imago mundi*, pp. 14–18.

Part I

An Introduction to the *Imago Mundi* tradition

1. *Understanding the World.* An Overview of the *Imago mundi*

Abstract: This is the first of two introductory overview chapters in this book. This chapter first introduces the concept of the medieval 'encyclopedia' and discusses the genre category to which the *Imago mundi* belongs. It then addresses the text's origin, composition, and the question of the identity of its author, Honorius Augustodunensis. The chapter concludes with a survey of the contents of the *Imago mundi*, focusing in particular on its geographical and astronomical Book I. This part of the encyclopedia is constructed according to the doctrine of the four elements: earth, water, air and fire. The section dedicated to the element of 'earth' is also quadripartite, covering Asia, Europe, Africa, and Islands (of the Mediterranean) in that order.

Keywords: medieval encyclopedias, *Imago mundi*, Honorius Augustodunensis; the four elements; T-O maps; *mappae mundi*

Much like Ruggiero on the hippogriff, we have the advantage of a bird's-eye view, and can take in the whole of the world described in the *Imago mundi*. Prior to discussing the text's contents, however, a few words are in order concerning its classification and its context of creation. This chapter, therefore, first discusses the label we use for the text (its 'genre'), then its origins, and finally its contents.

Medieval 'Encyclopedias'

'Every age has its peculiar encyclopaedia', according to Fowler, and in the Middle Ages the particular genre of geographical encyclopedia was very

Petrovskaia, N.I. *Transforming Europe in the Images of the World, 1110–1500. Fuzzy Geographies.* Taylor & Francis Group, 2025
DOI 10.5117/9789048563166_CH01

popular.[1] The 'encyclopedia' which generated the influential medieval tradition discussed in this book is the *Imago mundi* (also known as *De imagine mundi* 'On the Image of the World' or *De imagine mundi libri tres* 'Three Books on the Image of the World') of Honorius Augustodunensis.[2]

Although the *Imago mundi* and related texts are commonly referred to as 'encyclopaedias', as has been observed by the late great Jacques Le Goff, the word 'encyclopedia' itself is not attested in French or English until the sixteenth century, and its application to medieval works is anachronistic.[3] Although anachronistic, the label is descriptive and convenient for the purposes of modern classification. Indeed, the use of the term has been accepted as convention by most medievalists, it is also followed in this book.[4] We need to remember, however, that this 'genre' classification is

1 Robert L. Fowler, 'Encyclopaedias: Definitions and Theoretical Problems', in *Pre-Modern Encyclopaedic Texts Proceedings of the Second COMERS Congress, Groningen, 1–4 July 1996*, ed. by Peter Binkley (Leiden: Brill, 1997), pp. 3–29 (pp. 7, 18).

2 For references, see the Introduction, note 9.

3 Le Goff, 'Pourquoi le XIIIe siècle a-t-il été un siècle d'encyclopédisme?' in *L'enciclopedismo medievale*, ed. by M. Picone (Ravenna: Longo, 1994), pp. 23–40 (pp. 25–26). For instance, Bernard Ribémont classes the *Imago mundi* as an *encyclopédie éclatée*, an intermediate type between a *tractatus* (text with a very narrow focus) and an encyclopedia (text with a very broad focus); Ribémont, *Les origines des encyclopédies médiévales: D'Isidore de Seville aux Carolingiens* (Paris: Champion, 2001), pp. 234, 235–36 and Ribémont, 'L'encyclopedisme médiéval: de la définition d'un genre à son apogée. Sur la pertinence des notions d'apogée et de décadence', in Ribémont, *De Natura Rerum: Études sur les encyclopédies médiévales* (Orleans: Paradigme, 1995), pp. 11–68 (pp. 21–22 n. 14).

4 See, for instance, Arnaud Zucker, 'Introduction', in *Encyclopédire: Formes de L'Ambition Encyclopédique dans l'Antiquité et au Moyen Âge*, ed. by Zucker, Collection d'études médiévales de Nice 14 (Turnhout: Brepols, 2013), pp. 11–32 (p. 12); Bernard Ribémont, *La 'Renaissance' du XIIe siècle et l'encyclopédisme*, Essais sur le Moyen Âge 27 (Paris, 2002); Ribémont, 'On the Definition of an Encyclopedic Genre in the Middle Ages', in *Pre-Modern Encyclopaedic Texts*, ed. by Binkley, pp. 47–61; Benoît Beyer de Ryke, 'Le miroir du monde: un parcours dans l'encyclopédisme médiéval', *Revue belge de philologie et d'histoire* 81 (2003), 1243–1274; Picone, ed., *L'encyclopedismo medievale*; Isabelle Draelants, 'Le "siècle de l'encyclopédisme": conditions et critères de définition d'un genre', in *Encyclopédire*, ed. by Zucker, pp. 81–106 (p. 82); see also the essays in *Cahiers d'histoire mondiale* 9.3 (1966). See also the discussion in Mary Franklin-Brown, *Reading the World: Encyclopedic Writing in the Scholastic Age* (Chicago: Chicago University Press, 2012), pp. 9–10. A related question is whether the various texts we now group together as medieval 'encyclopedias' were perceived by contemporaries as belonging to the same genre; for discussions, see Ribémont, 'On the Definition of an Encyclopaedic Genre', p. 47, and Franklin-Brown, *Reading the World*, p. 9. The issue of the term's use for *Image du monde*, one of the members of the textual family examined in the present book, has been raised before; see Katherine A. Brown, 'The Vernacular Universe: Gossuin de Metz's *Image du monde, translatio studii*, and Vernacular Narrative', *Viator*, 44 (2013), 137–58 (p. 139) and Franklin-Brown, *Reading the World*, p. 9. See also the discussion of genres below, pp. 113–14.

modern, and therefore any apparent genre cross-fertilisation when *Imago mundi* is adapted into narrative, fictional texts, may be a product of modern classification practices rather than actual crossing of genre boundaries.

The common denominator in all these texts is a drive towards the systematisation of knowledge, or in the words of Isabelle Draelants, 'L'ambition encyclopédique, c'est-à-dire la position de l'esprit qui vise un système, est une perspective intellectuelle théorique qui organise les sciences et présente une synthèse des savoirs' ('The encyclopedic ambition, that is to say the mental attitude directed at systematisation, is an intellectual theoretical perspective which organises the sciences and presents a synthesis of knowledge').[5] Until about the middle of the thirteenth century, Latin works are at the core of this development.[6]

The *Imago mundi* family of texts is also characteristic of medieval encyclopedism in that, within this tradition, the turn towards the vernacular occurs, as it does elsewhere, in the middle of the thirteenth century.[7] We will continue to use the label 'encyclopedic', therefore, but as a characteristic rather than genre label. The characteristic can be shared across what we might now perceive as genre boundaries, and across translation over linguistic and geographical borders. The quotation from *Orlando Furioso* which opened this book also illustrates the impressive capacity of geographical information to ooze into new textual environments in medieval written culture, permeating literary production across what we would now describe as different genres.[8]

5 Isabelle Draelants, 'Le "siècle de l'encyclopédisme"', pp. 82–85. My translation.

6 As observed by David Napolitano, among others; 'Adjusting the Mirror: A Political Remake of Brunetto Latini's *Li Livres dou Tresor*', in *The Mirror in Medieval and Early Modern Culture; Specular Reflections*, ed. by Nancy M. Frelick (Turnhout: Brepols, 2016), pp. 89–111 (p. 89). It must be noted, however, that the *Imago mundi* is simpler in both matter and presentation than the 'core' works named by Napolitano: Thomas of Cantimpré, *De naturis rerum* (1237–40); Bartholomaeus Anglicus, *De proprietatibus rerum* (*c.* 1250); Vincent of Beauvais, *Speculum maius* (1244–59); Napolitano, 'Adjusting the Mirror', p. 89. Both Thomas of Cantimpré and Vincent of Beauvais may have used the *Imago mundi*; see Karl Schorbach, *Studien über das deutsche Volksbuch Lucidarius und seine Bearbeitungen in fremden Sprachen*, Quellen und Forschungen zur Sprach- und Kulturgeschichte der germanischen Völker, 74 (Strassburg: Karl J. Trübner, 1894), p. 162.

7 Cf. Napolitano, 'Adjusting the Mirror', p. 89.

8 A similar view lies at the base of Matthew Boyd Goldie, *Scribes of Space: Place in Middle English Literature and Late Medieval Science* (Ithaca, NY and London: Cornell University Press, 2019), which juxtaposes Middle English literary and scientific materials; for a description of the structuring principle of his book, see p. 3. Whereas the materials discussed by Goldie display, as he puts it, 'substantial, disciplinary, and formal differences between the science and the

This flexibility and fluidity of the tradition expresses also itself in the fluidity of attribution and the manner in which the text is referred to (or refers to itself). This terminological variation of references to this text in medieval and subsequently also modern discussions leads to a major difficulty in evaluating the full extent of the *Imago mundi* tradition and is another reason to keep using the label 'encyclopedia', approximate though it is. In the manuscripts, this text is occasionally labelled *Mappa mundi* or even *Speculum mundi*, and ascribed to assorted major medieval authorities, such as Isidore, Orosius, Bede, Henry of Huntingdon, and Anselm.[9] The ambiguity concerning the work's authorship is not limited to the medieval manuscript tradition. Misattributions continue well into the age of print.[10] In early modern and modern discussions, the text is variously referred to as *Imago mundi* and *De imagine mundi*, sometimes attributed to Honorius Augustodunensis, sometimes to Honorius of Autun, and sometimes designated as anonymous, with no reference to Honorius but with a reference to its medieval (mis-)attribution to Anselm of Canterbury.[11] This brings us to

literature', the *Imago mundi*–based texts discussed in the present book present a case where content transcends such differences.

9 Michael W. Twomey, 'Honorius Augustodunensis', *Trade, Travel, and Exploration in the Middle Ages: An Encyclopedia*, ed. by John Block Friedman and Kristen Mossler Frigg, with Scott D. Westrem and Gregory G. Guzman (New York and London: Routlledge, 2000), pp. 259–61 (p. 260). The misattribution to Isidore is by the Renaissance author Domenico Silvestri in *De insulis et earum proprietatibus*, who says that the text (cited as *De imagine mundi*) refers to Saint Brendan's visit to the island called *Perdita*; see quotation and discussion in Martínez, 'El mito de la isla perdida perdida y su tradición en la historia, cartografiía, literatura y arte', *Revista de filologia de la Universidad de La Laguna*, 16 (1998), 143–84 (pp. 154–57), in the context of further discussion of the association of the Brendan legend with the *Imago mundi* tradition, and references to two further Spanish texts in the tradition which are attributed to Isidore. See also below, p. 125 for discussion of the Brendan legend; and pp. 58, 63–65, 74, 98–99, 121–24, 129–35 for more on the Spanish translations of the *Imago mundi.*

10 See the brief overview in the Conclusion, p. 168.

11 The lack of reference to Honorius by name in modern discussions relating to texts which cite or adapt sections of the *Imago mundi* is often a function of the attribution practices in the period and area which is being discussed. For more on the misattribution to Anselm and its consequences for the *Nachleben* of Honorius's work, see Marie-Odile Garrigues, 'Honorius Augustodunensis et l'Italie', *Mélanges de l'École Française de Rome. Moyen Âge, Temps Modernes*, 84 (1952), 511–30 (p. 529). For more on the association between Honorius and Anselm, see Robert Darwin Crouse, 'Honorius Augustodunensis: Disciple of Anselm?', *Analecta Anselmiana*, 4 (1975), 131–39. The twentieth century also saw some debate on the subject of whether Honorius was indeed the author of the text. Three scholars have argued against Honorius's identification as the author of the *Imago mundi*: Johann von Kelle, P. Duhem (on the basis that manuscripts of the *Petite philosophie*, the Anglo-Norman translation of *Imago mundi*, circulated under the name of Henri or Honoré 'Angloy'; to him the first two books seemed to be of English origin, and Pliny the Elder and Isidore were thus known to the author

the subject of the text's author and compositional context, which we will address before turning to a brief overview of its contents.

About the Author and Concerning Context

After prolonged debate during the course of the twentieth century, the consensus on Honorius Augustodunensis appears to be that he was German, English, or Irish, and that Augustodunensis is not to be interpreted as Autun in France, but as a reference to *Augustodunum*, an Irish foundation near Regensburg in Germany.[12] If he was English or Irish, he had spent much of

only via Bede), and A. Cordiolani, who notes that English manuscripts attribute the text to one Henricus, whom he identifies as Henry of Huntingdon. Marie-Odile Garrigues, 'L'oeuvre d'Honorius Augustodunensis: Inventaire Critique' (1), *Abhandlungen der Braunschweigischen Wissenschaftlichen Gesellschaft*, 38 (1986), 7–136 (pp. 27–28); Cordiolani, 'Un manuscrit de comput et d'astronomie des XIIème-XIVème siècles, le manuscrit 467 de l'université de Glasgow de l'*Imago mundi*', *Scriptorium*, 3.1 (1949), 75–79; J. Von Kelle, 'Untersuchungen über den nicht nachweisbaren Honorius Augustodunensis presbyter und scholasticus und die ihm zugeschiebenen Werke', in *Sitzungsberichte des philosophisch-historischen Klasse der kaiserlichen Akademie der Wissenschaften zu Wien*, 102 (1905), fasc. 2; Duhem, *Le Système du monde, hisoire des doctrines cosmologiques de Platon à Copernic*, vol. III (Paris, 1919), pp. 24–31. As Sanford notes, the text features in a list of Honorius's works in the *Luminaribus ecclesiae*; 'Honorius *Presbyter* and *Scholasticus*', *Speculum*, 23 (1948), 397–425 (p. 397); with reference to *De Luminaribus ecclesiae* IV.17 in PL 172, ed. by J.-P. Migne (Paris, 1895), pp. 232–34. It is worth pointing out that the correspondence between the text of the *Imago mundi* and the *Etymologiae* is too close to postulate Bede as intermediary. Indeed, the texts are so close that in their analysis of the *Semeiança del mundo*, the text's modern editors mistakenly supposed it to have been based directly on Isidore, whereas in reality it is for the most part a verbatim translation of the *Imago mundi*; William E. Bull and Harry F. Williams, ed., *Semeiança del mundo: A Medieval Description of the World* (Berkeley and Los Angeles: University of California Press, 1959), p. 16; but see also Juan Casas Rigall, 'Razas humanas portentosas en las partidas remotas del mundo (de Benjamín de Tutela a Cristóbal Colón)', in *Maravillas, peregrinaciones y utopías: Literatura de viajes en el mundo románico*, ed. by Rafael Beltrán (Valencia: Universitat de Valencia, 2002), pp. 253–90 (p. 273).

12 Sanford, 'Honorius', pp. 398–99; Beyer de Ryke, 'Le miroir du monde', p. 1251 n. 33. Honorius is considered to be an Irishman in Garrigues, 'Honorius Augustodunensis et l'Italie', p. 511 and Roger E. Reynolds, 'Further Evidence for the Irish Origin of Honorius Augustodunensis', *Vivarium*, 7.1 (1969), 1–7. Two recent discussions, which I have not been able to consult, address Honorius's identity: Jacek Dębicki, *Zachodni portal katedry świętego Łazarza w Autun: Studium z historii sztuki i historii idei* (Kraków: Universitas, 2002), pp. 152–89 and Marek Jamróz, *Bóg, świat i człowiek w pismach Honoriusza Augustodunensisa* (Lublin: Wydawnictwo KUL, 2008). The references are from Leszek Wojciechowski, 'Geographical Descriptions of Spain in Latin Medieval Encyclopedias: From Isidore of Seville to Vincent of Beauvais (VII–XIII C.)', *Roczniki Humanistyczne*, 67 (2019), 31–55. For more on the discussions regarding authorship, see Garrigues, 'L'oeuvre d'Honorius Augustodunensis', pp. 27–28.

his time in Germany. If German, he appears to have travelled to Canterbury, to St Anselm, before settling down in Regensburg in the latter part of his life.[13] It will be seen in the following chapters that such a transnational carreer is not unusual in the *Imago mundi* tradition, and is a characteristic of both the writers and the texts they produced.

Valerie Flint distinguishes four redactions made by Honorius in 1110, 1123, 1133, and 1139 respectively.[14] Each version adds new information (sometimes as sentences, sometimes even chapters).[15] These changes sometimes help identify which version was used in vernacular translations or adaptations.[16] It is the earliest of these versions (1110), that Flint suggests may have been conceived if not composed in England, while later versions appear to be associated with Regensburg.[17]

It is frequently remarked that the *Imago mundi*, surviving in over a hundred medieval manuscripts, was enormously influential in the Middle Ages.[18] Not only was it translated into many European vernaculars but it also served as the main source for numerous later encyclopedic texts, and

13 For an overview, see Michael W. Twomey, 'Honorius Augustodunensis (c. 1080–c.1140)', in *Routledge Resources Online: Medieval Studies*, https://doi.org/10.4324/9780415791182-RMEO96-1, pp. 1–2. See also, however, the counter-argument for the association with Canterbury in Gleb Schmidt, 'From Manual to Best-Seller: The History of Honorius Augustodunensis's *Elucidarium*', in *Books of Knowledge in Late Medieval Europe: Circulation and Reception of Popular Texts*, ed. by Pavlína Cermanová and Václav Žůrek, Utrecht Studies in Medieval Literacy, 52 (Turnhout: Brepols, 2021), pp. 137–63 (pp. 140–41, esp. n. 5).

14 Honorius Augustodunensis, *Imago mundi*, ed. by Valerie I. J. Flint, *Archives d'histoire doctrinale et littéraire du Moyen Âge*, 49 (1982), p. 40 and Valerie I. J. Flint, 'World History in the Early Twelfth Century: The *Imago mundi* of Honorius Augustodunensis', in *The Writing of History in the Middle Ages: Essays Presented to Richard William Southern*, ed. by R. H. C. Davis and J. M. Wallace-Hadrill (Oxford: Clarendon Press, 1981), pp. 211–38 (p. 211). It is worth noting that Flint's discussion in the edition is suggestive of a holograph manuscript (or manuscripts) for which we have no evidence, in an echo, perhaps, of the so-called Old Philology which prioritised an authoritative manscript text, usually reconstructed and thus often largely fictional; see the discussion in Stephen D. Nichols, 'Introduction: Philology in a Manuscript Culture', *Speculum*, 65 (1990), 1–10 (p. 6). The exact relationship between the four versions of the *Imago mundi* identified by Flint, though worth revisiting in the future, does not affect the arguments in this book.

15 Flint, 'World History', pp. 216–20. See, however, discussion of transpositions in I.26 in Chapter 5.

16 See, for instance, the discussion of the Welsh translations, which derive from the two earliest versions of the text, in Natalia I. Petrovskaia, ed., *Delw y Byd: A Medieval Welsh Encyclopedia*, MHRA Library of Medieval Welsh Literature (London: MHRA, 2020), pp. 14–18, and this book, pp. 58, 60, 65, 96–98.

17 Flint, ed., *Imago mundi*, pp. 8–14; Flint, 'World History', pp. 211–12.

18 See, for instance, Graeme Dunphy, 'Historical Writing in and after the Old High German Period', in *German Literature of the Early Middle Ages*, ed. by Brian Murdoch (Woodbridge:

extracts were used in both Latin and vernacular works of devotional and literary as well as of encyclopedic nature.[19] It is one of two of Honorius's works that spawned a wide and influential tradition.

The other is the *Elucidarium*.[20] Vernacular versions of the *Elucidarium* include the French *La lumiere as lais*, as well as Spanish, Welsh, and Norse translations.[21] The reason for the popularity of these texts may be that the purpose of both was to provide easily accessible knowledge.[22] As Joëlle Ducos notes, while the *Elucidarium* was translated in its entirety, the *Imago mundi* is always only partially used for translations.[23] It is always

Boydell & Brewer; New York: Camden House, 2004), pp. 201–26 (p. 210); Flint, ed., *Imago mundi*, p. 18.

19 See pp. 59–79 below.

20 Yves Lefèvre, *'L'Elucidarium' et les lucidaires: Contribution, par l'histoire d'un texte, à l'histoire des croyances religieuses en France au Moyen Âge*, Bibliothèque des Écoles françaises d'Athènes et de Rome, 180 (Paris: E. de Brocard, 1954), pp. 343–497. For more, see Schmidt, 'From Manual to Best-Seller'.

21 For discussions of the text, see Lefèvre, *'L'Elucidarium'*; Ernstpeter Ruhe, *Elucidarium und Lucidaires: zur Rezeption des Werks von Honorius Augustodunensis in der Romania und in England*, Wissensliteratur im Mittelalter, 7 (Wiesbaden: Reichert, 1993); Martha Kleinhans, *'Lucidere vault tant a dire comme donnant lumiere': Untersuchung und Edition der Prosaversionen 2, 4 und 5 des 'Elucidarium*, Beihefte zur Zeitschrift für romanische Philologie, 248 (Tübingen, Niemeyer: 1993); Richard Kinkade, ed., *Los Lucidarios españoles: Estudio y edición* (Madrid: Gredos, 1968); Ana M. Montero, 'El *Lucidario* de Sancho IV: Redefinición de su relación textual con el *Elucidarius* de Honorius Augustodunensis y el *Setenario* de Alfonso X', in *The Theory and Practice of Translation in the Middle Ages*, ed. by R. Voaden, R. Tixier, T. Sanchez Roura, and J. R. Rytting (Turnhout: Brepols, 2003), pp. 49–59; *'Elucidarius' in Old Norse Translation*, ed. by Evelyn Scherabon Firchow and Kaaren Grimstad (Reykjavik: Stofnun Árna Magnússonar, 1989) and *The Old Norse 'Elucidarius': Original Text and English Translation*, ed. and trans. by Evelyn Scherabon Firchow, Medieval Texts and Translations (Columbia, SC: Camden House, 1992). Svanhildur Óskarsdóttir, 'Prose of Christian Instruction Instruction', in *A Companion to Old Norse-Icelandic Literature and Culture*, ed. by Rory McTurk, Blackwell Companions to Literature and Culture, 31 (Malden, MA, Oxford and Victoria: Blackwell, 2005), pp. 338–53 (p. 342); Rudolf Simek, *Heaven and Earth in the Middle Ages: The Physical World Before Columbus*, trans. by Angela Hall (Woodbridge: Boydell, 1996). Note that the German *Lucidarius* (compiled *c.* 1190–1195) has excerpts from the *Elucidarium*, the *Imago mundi*, and the *Gemma animae*; Sanford, 'Honorius', p. 398.

22 See, for instance, Flint, 'World History', p. 212. The apparent simplicity of Honorius's exposition has been much commented on. See, for instance, Flint, ed., *Imago mundi*, p. 14; for further references, see above, p. 16 n. 10. According to Evelyn Edson, the *Imago mundi* may have been intended specifically for isolated religious communities lacking access to large libraries; Edson, *Mapping Time and Space: How Medieval Mapmakers Viewed their World*, The British Library Studies in Map History, 1(London: BL, 1997), p. 111.

23 Joëlle Ducos 'Que traduire en français? Traductions uniques et traductions multiples', in *Translation and Authority: Authorities in Translation*, ed. by Pieter de Leemans and Michèle Goyens, The Medieval Translator, 16 (Turnhout: Brepols, 2017), pp. 39–52 (p. 41).

Book I of the *Imago mundi* (comprising the description of the physical world) that is used, and therefore one could adjust the comparison to measure the spread of the *Elucidarium* against that of *Imago mundi* I. As a compendium of established knowledge, the information contained in the latter is entirely derivative and belongs to what was by then a well-established Western encyclopedic tradition. This combination of thoroughness with accessible form and uncomplicated Latin partially accounts for the immense success of Honorius's encyclopedic work. The flexibility of the geographical framework he offers, described and discussed further below in terms of archaism and fuzzy sets, is the other major factor.

The *Imago mundi* is preceded by two letters in lieu of an authorial preface. The letters, one to and one from the author, describe the origin and purpose of the text.[24] Although this might appear to set up a dialogue format used by Honorius himself in the *Elucidarium*, it is not maintained in the *Imago mundi*. Within the manuscript tradition, the addressee of Honorius's response, and thus dedicatee of the work, varies between Christianus and Henricus.[25] Christianus has been identified as the abbot of the Schottenkloster of St James at Regensburg, and the various possible candidates for 'Henricus' include Henry the Black (1072–1126) of Bavaria, Henry of Blois (*c.* 1096–1171), abbot of Glastonbury and bishop of Winchester, and finally, Henry of Huntingdon (*c.* 1088–*c.* 1157).[26] Flint suggests that the dedicatee of the work is to be identified with the latter.[27] This association may well explain the occasional attribution of the work to Henry of Huntingdon in medieval manuscripts. The uncertainty surrounding the dedication and patronage of the work reflects the uncertainty surrounding the circumstances and location of its composition.

24 Flint, ed., *Imago Mundi*, pp. 48–49. For a detailed discussion see Garrigues, 'L'oeuvre', pp. 32–33. The letters are not always reproduced in the manuscript tradition. They are missing, for instance, from Augsburg, Staats- und Stadtbibliothek Augsburg 2 Cod. 108 (s. xv, Eichstatt?) (Figure 10 in this book), and from one of the versions of the Welsh *Delw y Byd* (Version B); Petrovskaia, ed., *Delw y Byd*. The possibility that these represent a separate tradition, which circulated *sans* introductory letters, deserves a separate study. For a full chapter list of the *Imago mundi*, see Appendix I, pp. 173–75.

25 Flint, ed., *Imago Mundi*, p. 10.

26 Valerie I. J. Flint, 'Honorius Augustodunensis', in *Authors of the Middle Ages II: Historical and Religious Writers of the Latin West*, Nos 5–6, ed. by Patrick J. Geary (Aldershot: Variorum Reprints, 1995), pp. 89–183; Twomey, 'Honorius', p. 259; Flint, ed., *Imago Mundi*, pp. 10–12.

27 Flint, ed., *Imago Mundi*, p. 12.

The Contents of the *Imago Mundi*

After the opening letters, the structure of the *Imago mundi* is tripartite. The three Books describe space, time, and the events taking place in space and time.[28]

The work connects the now disparate disciplines of history and geography in linking space-time into a single system.[29] This synthesis of space-time is also visible in Book I when it is taken on its own, as it almost invariably was in the vernacular translations and adaptations. It is structured according to the medieval doctrine of the four elements, conceived as concentric spheres encompassing all creation. Earth (geography), Water (oceans, rivers, weather phenomena), Air (winds), Fire (heavenly spheres, astronomy, the zodiac) are treated in order.

The first, geographical, section of the text presents a tripartite world. This tripartite division of the inhabited world is generally acknowledged to derive from ancient Greek geography.[30] It reached the *Imago mundi* via Isidore's *Etymologiae*: *Divisus est autem trifarie, e quibus una pars Asia, altera Europa, tertia Africa nuncupatur* ('It is divided into three parts, one of which is called Asia, the second Europe, the third Africa').[31]

28 It is this structure that provides inspiration for the structure of Part II of this book, in which Chapter 3 is Time, Chapter 4 is Space, and Chapter 5 is Movement. Flint, ed., *Imago Mundi*, p. 13; Flint, 'World History', pp. 212–13; Twomey, 'Honorius', p. 260; Edson, *Mapping Time and Space*, pp. 112–13.

29 See discussion of the *chronotope* in the Introduction.

30 Martin W. Lewis and Karen E. Wigen, *The Myth of Continents: A Critique of Metageography* (Berkeley: University of California Press, 1997), p. 22. In the earlier period, it appears to have coexisted with a dualist division of Europe–Asia, and rose to dominance only in Late Antiquity. This comparatively minor tradition of distinguishing two, rather than three landmasses was exemplified by Hektaios of Milet and by Isocrates in *Panegyrikos*, 179; Isocrates, *Isocrates*, 1, with translation by George Norlin, Loeb Classical Library, 209 (Cambridge MA and London: Harvard University Press, 1945), pp. 234–35). It appears to have been still known to Paulus Orosius and St Augustine. Baumgärtner, 'Europa in der Kartographie des Mittelalters', pp. 11–14; Alfred Stückelberger, 'Das Europabild bei Ptolemaios', in *Europa im Weltbild des Mittelalters. Kartographische Konzepte*, ed. By Ingrid Baumgärtner and Hartmut Kugler, Orbis mediaevalis, 10 (Berlin: Akademie Verlad, 2008), pp. 31–44 (p. 31 n. 4). The fluidity of the Europe–Africa boundary which we find in the *Image du monde* may be read in the context of this variation; see below, pp. 111–21.

31 *Etym.*, XIV.2; W. M. Lindsay, ed., *Isidori Hispalensis Episcopi Etymologiarvm sive Originvm Libri XX*, vol. 2 (Oxford: Oxford University Press, 2011); trans. in Stephen A. Barney, W. J. Lewis, J. A. Beach, and Oliver Berghof, trans., *The Etymologies of Isidore of Seville* (Cambridge: Cambridge University Press, 2006), p. 285. While, as has been mentioned above, the tripartite view derives from classical geographical traditions, but in the Middle Ages the knowledge of it is primarily derived from Isidore; David Woodward, 'Medieval *Mappaemundi*' in *History of Cartography I:*

The tripartite world structure is laid out in I.7, following the chapters on the four elements, the spherical shape of the world, and the names of the earth: *Habitabilis zona que a nobis incolitur, in tres partes Mediterraneo mari dirimitur, quarum una Asia, altera Europa, tercia Affrica dicitur* ('The habitable zone which is occupied by us is divided into three parts by the Mediterranean sea, of which one is called Asia, another Europe, and the third Africa').[32] This tripartite formula, the dominant way of presenting the inhabited world in the Middle Ages, is shared by all the works of the *Imago mundi* tradition.[33] The order of the regions in I.7, corresponds to the

Cartography in Prehistoric, Ancient, and Medieval Europe and the Mediterranean, ed. by J. B. Harley and David Woodward (Chicago and London: University of Chicago Press, 1987), pp. 286–370 (pp. 301–02).

32 Flint, ed., *Imago mundi*, p. 52. Translations from this text in this book are mine unless otherwise noted.

33 *Imago mundi* I does not associate the three parts of the world with the Sons of Noah. They are referred to but without that association in I.13 and 15; Flint, ed., *Imago mundi*, pp. 55–56. He refers to them again in Book III and in *Gemma animae*; Oschema, *Bilder von Europa*, pp. 338–39; Honorius Augustodunensis, *Gemma animae* in PL 172, col. 627. See also Caroline D. Eckhardt, 'One Third of the Earth? Europe Seen and Unseen in the Middle English Chronicles of the Fourteenth Century', *Comparative Literature*, 58.4 'The Idea of Europe' (2006), 313–38 (p. 323). Honorius's reference to the Sons of Noah in *Imago mundi* Book III appears to have found an echo in Scandinavian literature; see Thomas D. Hill, '*Rígsþula*: Some Medieval Christian Analogues', *Speculum*, 61 (1986), 79–89. For the relation between the three parts of the world and the Sons of Noah, see, for instance, David Woodward, 'Reality, Symbolism, Time and Space in Medieval World Maps', *Annals of the Association of American Geographers*, 75 (1985), 510–21; Benjamin Braude, 'The Sons of Noah and the Construction of Ethnic and Geographical Identities in the Medieval and Early Modern Periods', *The William and Mary Quarterly*, 54 (1997), 103–42; Suzanne Conklin Akbari, *Idols in the East: European Representations of Islam and the Orient, 1100–1450* (Ithaca, NY and London: Cornell University Press, 2009), pp. 40–42, 137–38; Marcia Kupfer, 'The Noachide Dispersion in English Mappae Mundi, ca. 960–ca. 1130', *Peregrinations: Journal of Medieval Art and Architecture*, 4 (2013), 81–106; and the brief discussions in James S. Romm, 'Continents, Climates, and Cultures: Greek Theories of Global Structure', in *Geography and Ethnography: Perceptions of the World in Pre-Modern Societies*, ed. by Kurt A. Raaflaub and Richard J. A. Talbert (Chichester: Wiley–Blackwell, 2010), pp. 215–35 (pp. 228–31) and from Wojciechowski, 'Geographical Descriptions of Spain in Latin Medieval Encyclopedias', pp. 39–40; Aníbal A. Biglieri, *Las ideas geográficas y la imagen del mundo en la literatura española medieval*, Medievalia Hispanica, 17 (Madrid: Iberoamericana; Frankfurt am Main: Vervuert, 2012), pp. 68–72. Reference to the sons of Noah is absent from Honorius's main source, Isidore's *Etymologiae*. For a discussion of the importance of this absence in the Isidoran tradition, see Chet Van Duzer, and Sandra Sáenz-López Pérez, '*Tres filii Noe diviserunt orbem post diluvium*: The World Map in British Library Add. MS 37049', *Word & Image*, 26 (2009), 21–39 (p. 32). The apparent need to add the Noachian tripartition to the text, pointed out by Van Duzer and Sáenz-López Pérez in relation to the medieval illustrators of Isidore's text, appears not to have been felt by Honorius; Van Duzer and Sáenz-López Pérez, '*Tres filii Noe*', p. 32 and p. 39 n. 88. They point out that the Noachide reference is also absent in the *Image du monde* but without

order of their discussion: Asia, Europe, Africa. It is associated specifically with encyclopedic genres.[34] Texts in the historical genres, by contrast, tend to follow rather the order Asia–Africa–Europe.[35] This use appears linked to the notion of *translatio imperii* and to the spread of Trojan origin legends, as Aeneas's trajectory took him from Troy to Rome via Carthage (Asia→Africa→Europe).[36]

The tripartite description of the world in *Imago mundi* corresponds to the worldview transmitted through the medieval world maps, the *mappae mundi* (Latin, singular: *mappa mundi*), particularly of the T-O type.[37] An important distinction between the visual diagrammatic depiction of world geography in the T-O schema and the textual map, however, needs to be

noting the relationship between that text and Isidore (ibid.). One possible explanation for the exclusion of any reference to Noah's sons in *Imago mundi* I is that Honorius may have wished to focus on a purely physical description of the geographical world, without engaging with the various implications of the biblical allusion. In support of this one might point out that he also does not refer to Adam and Eve when discussing Paradise (I.8). References to the sons of Noah are reincorporated into the geographical description by many of the texts that rely on the *Imago mundi*, however, possibly as a reflection of precisely that need. Gervase of Tilbury, for instance, introduces his description of the tripartite world in *Otia imperialia* with an explicit reference; *Otia Imperialia* II.1, in Gervase of Tilbury, *Otia Imperialia: Recreation for an Emperor*, ed. and trans. by S. E. Banks and J. W. Binns (Oxford: Clarendon Press, 2002), pp. 176–77.

34 Natalia I. Petrovskaia, 'La formulation tripartite du monde dans les encyclopédies et textes littéraires du Moyen Âge', in *La Formule au Moyen Âge IV / Formulas in Medieval Culture IV*, ed. by E. Louviot, C. Garcia, and S. Morrison, ARTeM Atelier de Recherche sur les Textes Médiévaux 31 (Turnhout: Brepols, 2022), pp. 197–217. In the *Imago mundi* tradition, even those texts which cross genre boundaries (as we understand them), such as *Alexanders geesten* and *Laberinto de fortuna*, retain the order of description of the Latin.

35 There may be exceptions to the trend, and the question bears revisiting; see Mauntel, *Erdteile*, pp. 259–62. I am grateful to Christoph Mauntel for this reference.

36 Petrovskaia, 'La formulation tripartite' and this book, p. 73, 137–38.

37 This type of map represented the inhabited part of the world in a circle, with the East at the top, and the three parts (Asia, Europe, and Africa) demarcated by the Mediterranean, the Don, and the Nile. Whether the Mediterranean would have been seen as a divider or a connector in the Middle Ages is a point further discussed below, pp. 106–07, 126–27. Its connecting power was first forcefully argued by Fernand Braudel in *La Méditerranée et le Monde Méditerranéen à l'Epoque de Philippe II* (Paris: A. Colin, 1949). For a more recent discussion of the role of maritime routes in mediating cultural contact and exchange on a global scale see Hyunhee Park, *Mapping the Chinese and Islamic Worlds* (Cambridge: Cambridge University Press, 2015), pp. 21, 29–34, 42–54. The bibliography on the *mappae mundi* is vast. Useful starting points are: Woodward, 'Medieval *Mappaemundi*'; P. D. A. Harvey, *Medieval Maps* (Toronto: University of Toronto Press, 1991); Evelyn Edson, *The World Map, 1300–1492: The Persistence of Tradition and Transformation* (Baltimore: Johns Hopkins University Press, 2007); Evelyn Edson, 'The Medieval World View: Contemplating the Mappamundi', *History Compass*, 8 (2010), 503–17. Still useful is the catalogue of Marcel Destombes, *Mappaemondes A.D. 1200–1500: Catalogue préparé par la commission des cartes anciennes de l'Union géographique internationale* (Amsterdam: N. Israel, 1964).

highlighted. T-O diagrams present the world's tripartite structure with boundaries between the three parts of the world clearly demarcated with lines. This creates an impression of a conception of stable linear boundaries. However, it would be misleading to think of the world map in the textual tradition of the *Imago mundi* and its descendants in terms of such diagrammatically defined areas. I will show below that unlike the diagrams (and to some extent, arguably, the related *mappae mundi*), in the *Imago mundi* family, each part of the world represents a fuzzy set, with flexible areas of overlap between them.

A further important distinction between the diagrammatic T-O tradition and the *Imago mundi* is that in the text a fourth subsection is added to this tripartite geographic scheme. This forth subsection is devoted to islands, specifically those located in the Mediterranean, and thus essentially corresponds to the T of the T-O map.[38] It includes Sicily, Sardinia, Corsica, and Ibiza, among others. The Mediterranean therefore becomes a fuzzy set in itself rather than a boundary separating Europe and Africa. That the Mediterranean, and the territories on the landmasses surrounding it, could be seen either as a separate set or as members of one or other of the three parts of the world, depending on time and place of writing, is a topic explored further in Chapter 4. This apparent flexibility is discussed in terms of degrees of membership of fuzzy sets. The subsection in the *Imago mundi* which follows this is dedicated to the subject of Hell, which is perceived as contained underground.[39] This concludes the five subsections of the description of Earth.

The section of the *Imago mundi* devoted to Europe (I.21–29) follows the discussion of Asia (I.8–I.20) and begins with the words: *Europa ab occidente usque ad septentrionem* [...] *extenditur* ('Europe extends from the west to the north').[40] I.21 describes Europe as a whole, and is illustrative of information the text provides about the various regions it covers. Typical of the text, for instance, are the etymological explanations, often given, as in I.21, with alternatives. The description here and throughout the text is

38 This corresponds to chapters 33–35 of the *Imago mundi*; Flint, ed., *Imago mundi*, pp. 64–66. See discussion below, pp. 123–28. The fact that islands are separated out suggests, on the one hand, that they occupy a liminal space between the three parts of the world, and on the other echoes the isolation of islands as spaces; see the studies published in Maeve McCusker and Anthony Soares, eds, *Islanded Identities: Constructions of Postcolonial Cultural Insularity* (Amsterdam and New York: Rodopi, 2011).

39 Flint, ed., *Imago mundi*, pp. 66–67.

40 Flint, ed., *Imago mundi*, pp. 52–59 and 59–63; Flint, ed., *Imago mundi* I.7 (p. 7). 'Europe extends from the west to the north' (my translation).

cumulative. Rather than focusing on borders and boundaries – that is, on lines that divide – the text describes areas by presenting lists of smaller geographical units contained therein.[41] Simply put, it presents regions as sets. The technical definition for an entity that contains within it smaller entities, and is defined by and comprised of these smaller entities, is a 'set'.[42]

Although the rest of the *Imago mundi* will not have a major role to play in the present book, a brief overview is provided here for the sake of context. The following part of Book I is dedicated to the element of Water.[43] This section deals with the concepts of the oceans, fresh and saltwater, fish inhabiting the sea, and tempests. This is followed by the Air section, which discusses the twelve winds, primarily based on the wind-rose of Isidore of Seville, as well as aerial phenomena such as the rainbow, rain, fog, and the clouds.[44] Finally, the section devoted to the element of Fire covers

41 In terms of both structure and the language used, the same approach, which she describes as 'enumerative and cumulative', has been observed by Nathalie Bouloux in a much later text, the fifteenth-century *Le livre de la description des pays* by Gilles le Bouvier; 'From Gaul to the Kingdom of France: Representations of French Space in the Geographical Texts of the Middle Ages (Twelfth–Fifteenth Centuries)', in *Space in the Medieval West. Places, Territories and Imagines Communities*, ed. by Meredith Cohen and Fanny Madeline (Farnham and Burlington: Ashgate, 2014), pp. 197–217 (p. 209). A similar view is expressed by Robert Fawtier, who argues that the medieval notion of frontier was not of a line but rather a 'no man's land', and that as a consequence, a king of France, for instance, would not have had a clear view of the boundaries of a collection of lands he ruled. It is his description of the king's territory as 'l'ensemble des pays où il possédait des terres et des droits' that is relevant here; see Fawtier, 'Comment le roi de France, au début du XIVe siècle, pouvait-il se représenter son royaume' in *Mélanges offerts à M. Paul-Edmond Martin* (Geneva: Comité des mélanges P.-E. Martin, 1961), pp. 65–77 (p. 68). In the case of Europe, this can be illustrated by listing the chapters this section contains: 21 *De Europa* ('Of Europe'), 22 *De Scithia* ('Of Scythia'), 23 *De Germania Superior* ('Of Upper Germany'), 24 *De Germania Inferiore* ('Of Lower Germany'), 25 *De Grecia* ('Of Greece'), 26 *De Italia* ('Of Italy'), 27 *Gallia* ('Of Gaul'), 28 *Hispania* ('Of Spain'), 29 *Britannia* ('Britain'); *Imago mundi*, ed. Flint, pp. 59–63. Although these chapter divisions are cited here as they stand in the edition, and are the result of editorial choices made on the basis of a selection of manuscript witnesses, Flint's chapter divisions and headings – used in the present discussion – are primarily based on one of the manuscripts, Munchen, Bayerische Staatsbibliothek, clm. 536 (Prül, s. XII); ibid., pp. 25–26, 45. These headings are, however, representative, since most manuscripts of *Imago mundi* contain these as marginal annotations. Both the manuscript headings on which they are based and the structure of the chapters themselves attest to a cumulative pattern, an enumeration of neighbouring regions and their characteristics, without any significant demarcation of borders. For difficulties in identifying chapter breaks and a discussion of alternative possibilities of chapter division, based on choices made by the medieval translators of the text, see Petrovskaia, ed., *Delw y Byd*, pp. 29–30.

42 See discussion the introduction above, pp. 30–31.

43 Flint, ed., *Imago mundi*, pp. 67–72.

44 Flint, ed., *Imago mundi*, pp. 72–75.

astronomy, according to the Ptolemaic (geocentric) system, culminating in the description of the heaven of heavens.[45]

Book II concerns the measurement of time.[46] It discusses night and day, eclipses, the months of the year, the four seasons, planetary cycles, equinoxes and solstices, and other computistical concepts. Some of the concepts described are ecclesiastical in nature, such as the epact.[47]

Book III is the shortest and appears to be partly an epitome of Honorius's own *Summa Totius*.[48] It is a chronicle of universal history, structured according to the doctrine of the six ages of the world, commencing with the Creation and moving through the history of humanity as recounted in the Old Testament.[49] Its concluding section is dedicated to the history of Christianity, from the Roman to the German Empires. Its latter part is organised as a sequence of persecutions of Christians.

The structure of the *Imago mundi* reflects throughout what Jason Baxter has characterised as a 'concern with the principles of order' combined with 'a keen desire to capture a totalizing representation (as opposed to simply providing treasuries of natural knowledge to aid with exegesis)' which

45 Flint, ed., *Imago mundi*, pp. 75–92.

46 Flint, ed., *Imago mundi*, pp. 92–123.

47 For the relevance of astronomical concepts, observations, and calculations to ecclesiastical matters, see the recent discussion in Seb Falk, *The Light Ages: A Medieval Journey of Discovery* (Milton Keynes: Allen Lane, 2020). An epact is the measurement of time difference between the lunar and solar calendars and is crucial for calculating the date of Easter; see Stephen C. McCluskey, *Astronomies and Cultures in Early Medieval Europe* (Cambridge: Cambridge University Press, 1998), pp. 81–84; also Dale Kedwards, *The Mappae Mundi of Medieval Iceland* (Woodbridge: Boydell & Brewer, 2020), pp. 56, 58.

48 Flint, ed., *Imago mundi*, pp. 123–51; Flint, 'World History', pp. 214–15. The *Summa Totius* itself appears not to have been very popular, if manuscript evidence is taken as a guide, for it survives in only one incomplete thirteenth-century manuscript: Vienna Österreichische Nationalbibliothek MS 382 (Lambach, 1150–1199); ibid., p. 215 n. 3. For the date and description of the manuscript, see the online catalogue of the Österreichische Narionalbibliothek at http://data.onb.ac.at/rec/AC13951594 (accessed 28 March 2020). For the text, see D. R. Wilmans, ed., *Ex Honorii Augustodunensis Summa Totius et Imagine Mundi*, Monumenta Germaniae Historica. Scriptorum, 10 (Hannover, 1852), or J.-P. Migne, ed., PL 172, cols. 187–96.

49 Flint, 'World History', p. 213; Edson, *Mapping Time and Space*, pp. 26, 98–99, 100. See also Natalia I. Petrovskaia, '*Translatio* and Translation: The Duality of the Concept from the Middle Ages to the Early Modern Period', *Doshisha Studies in English*, 99 (2018), 119–36 (pp. 123–24), for further references. The primary sources of this section of the text appear to be the Book of Genesis, the Book of Numbers, Orosius, Eusebius-Jerome, Isidore, Cassiodorus, Bede, the *Historia Miscella*, Frutolf of Michelsberg, and, for the final sections, Hermannus Contractus and possibly an early version of the *Chronicon Wirziburgense* ('Chronicle of Würzburg'); see Flint, 'World History', p. 213; for the *Chronicon Wirziburgense*, see Stephen Penn, 'Chronicon Wirziburgense', in *Encyclopedia of the Medieval Chronicle*, ed. by Graeme Dunphy and Cristian Bratu, DOI: 10.1163/2213-2139_emc_SIM_00673 (accessed 11 April 2024).

distinguishes twelfth-century encyclopedic endeavours from earlier works, such as that of Isidore or Bede.[50] Indeed, the organisation of Honorius's text is radically different from both Isidore's *Etymologiae* and Bede's *De natura rerum*, which are his primary sources. The order in which the physical world is described by Isidore and by Bede is top-to-bottom: from God, through human beings, to inanimate things; by contrast, Honorius's Book I, as we have seen, starts with the earth and culminates in the description of the Heaven of Heavens.[51] It goes up. The re-ordering and re-organisation of material, shifting emphasis in shifting perspectives, is a theme that will resurface later in our discussion, in relation to the transformations of Europe in the text's different adaptations. In order to discuss these changes, we first need to look at what these adaptations are and how they might be described and classified.

50 Jason Baxter, '"Videmus nunc per speculum": The Mysticism and Naturalism of the Twelfth-Century *imago mundi*', *Haskins Society Journal*, 28 (2017), 119–42 (p. 126); citing Christel Meier-Staubach, 'Organisation of Knowledge and Encyclopaedic Ordo: Functions and Purposes of a Universal Literary Genre', in *Pre-Modern Encyclopaedic Texts*, ed. by Binkley, pp. 101–80.

51 See descriptions of the structure of Isidore's and Bede's works in Barney et al., 'Introduction' to *The Etymologies of Isidore of Seville*, p. 20, and Bede, *On the Nature of Things and On Times*, trans. by Calvin B. Kendall and Faith Wallis, Translated Texts for Historians, 56 (Liverpool: Liverpool University Press, 2010), pp. 3, 138. The physical world, as described in *Etymologiae*, Books XI–XVI, is ordered from humans to the earth, including materials.

2. *Translating Knowledge.* An Introduction to the *Imago Mundi* Family

Abstract: This chapter provides an overview of the multilingual tradition based on the *Imago mundi*. The major vernacular translations and adaptations of the Latin text are introduced here, including the the anonymous *Semeiança del mundo* and *Delw y Byd*, Gossouin de Metz's *Image du Monde*, Brunetto Latini's *Trésor*, and Fazio degli Uberti's *Dittamondo*, among others. These are grouped by type of adaptation under the following headings: 'Vernacular translations of *Imago mundi*'; 'Adaptations of *Imago mundi* and insertion of translated extracts in encyclopedic contexts'; 'Adaptations and translated extracts drawn into other genres'; 'Adaptations into Latin'.

Keywords: *translatio studii*; *Imago mundi*; genre; medieval encyclopedias; medieval translation; multilingual transmission

How do we get from the *Imago mundi* to *Orlando Furioso*? As mentioned above, the latter text is not a direct descendant of Honorius's encyclopedia. It does, however, share a common tradition, because the work to which it is a sequel, Boiardo's *L'Orlando Innamorato*, draws on Fazio degli Uberti's *Dittamondo*, which is part of the *Imago mundi* family.[1] The *Dittamondo* is perhaps the most distant of the texts drawing on the *Imago mundi* which will be discussed in this book. Most of its fellow travellers, however, are not usually discussed together, hence the need for this chapter. Its primary objective

1 For the text, see Fazio degli Uberti, *Il Dittamondo e le rime*, ed. by Giuseppe Corsi (Bari: Giuseppe Laterza & Figli, 1952), vol. 1, *Il Dittamondo*. For a recent discussion and up-to-date bibliography, see Juan Carlos D'Amico, 'Fazio degli Uberti, il Dittamondo e l'ideologia imperiale', in *Emperors and Imperial Discourse in Italy, c. 1300–1500: New Perspectives*, ed. by Anne Huijbers (Rome: École Française de Rome, 2022), pp. 89–104. The *Dittamondo* is further discussed below, pp. 75, 140–44.

Petrovskaia, N.I. *Transforming Europe in the Images of the World, 1110–1500. Fuzzy Geographies.* Taylor & Francis Group, 2025
DOI 10.5117/9789048563166_CH02

is to provide a general introduction to the translations and adaptations of the *Imago mundi*. I begin, as in Chapter 1, by discussing terminology. I then introduce a number of categories into which the texts to be discussed can be usefully grouped, an then discuss the texts within each category in order. Within each category the order of discussion is primarily chronological, insofar as the approximate dating of some of these works permits.

Although it is possible to use the term 'translation' very loosely to cover all types of use of *Imago mundi* material in new texts and contexts discussed in this book, it is more useful to describe the processes of adaptation undergone by this material in each case as 'rewriting'. I use the term according to the definition provided by Liedeke Plate and Els Rose: 'the process and product of the act of writing again', which 'encompasses a variety of activities, including the copying of a manuscript, the retelling of a story, the re-narration of an event, and the re-edition of a text'.[2] Crucially for the purposes of the present analysis it is also 'an act of transfer enabling cultural remembrance, rewriting inscribes time and difference'.[3] While Plate and Rose see rewriting as a 'technology of cultural memory' separate from translation and exegesis, the definition above can be stretched a little to encompass the processes and products of the *Imago mundi* tradition.[4] We can describe these texts as 'rewritings' of the Latin original, although there are significant differences between them. Some are close to what we would call 'translation', others are translated extracts or paraphrases, and still others are adaptations of extracts used in new contexts.

The ultimate cause for the difficulties in classifying these rewritings is not so much the lack of 'faithfulness' to the original as the differences between modern and medieval ideas about transmission. As Roger Ellis points out, our current understanding of translation as a process of cultural appropriation from perceived cultural superiors taking place between languages, was both inter- and intra-lingual in the medieval period.[5] This medieval idea of

2 The concept of rewriting is used here in the definition provided by Liedeke Plate and Els Rose in 'Rewriting, A Literary Concept for the Study of Cultural Memory: Towards a Transhistorical Approach to Cultural Remembrance', *Neophilologus*, 97 (2013), 611–25 (pp. 613–14). For a discussion of the criticism surrounding the metaphoric use of 'translation', see Susan Bassnett, 'From Cultural Turn to Translational Turn: a Transnational Journey', in *Literature, Geography, Translation: Studies in World Writing*, ed. by Cecilia Alvstad, Stefan Helgesson, and David Watson (Newcastle upon Tyne: Cambridge Scholars Publishing, 2011), pp. 67–80 (p. 74), citing concerns raised by Harish Trivedi.

3 Plate and Rose, 'Rewriting', p. 613.

4 Plate and Rose, 'Rewriting', p. 614.

5 Roger Ellis, 'Translation and Frontiers in Late Medieval England: Caxton, Kempe, and Mandeville', in *Frontiers in the Middle Ages: Proceedings of the Third European Congress of Medieval*

translation is therefore closer to transmission and thus the medieval notion of *translatio*.[6] Some of the *Imago mundi* texts identify their own position within this process of translation and appropriation, and their relation to the authority of the source(s) being appropriated, specifically within the framework of the *translatio studii*.[7]

Variants of the word 'translation' are used in the opening of both the *Image du monde*, 'translatez de latin en rommanz' ('translated from Latin into romance') and of *Trésor*, 'le quels translata maistre Brunet Latin' ('which master Brunet Latin translated'), but each text in fact relies on, rather than translates, their respective Latin and the French exemplars.[8] This is a case of rewriting. Thus, the use of 'translation'-related terms such as *translater* in our corpus raises the question of how these texts were understood by their authors and audiences.

The medieval concept of translation encompassed both linguistic transfer and movement, and it may be judicious to avoid making the distinction (required by modern terminology) between the two.[9] This echoes the connection, which can also be interpreted as identity, between the translation

Studies (Jyväskylä, 10–14 June 2003), ed. by O. Merisalo and P. Pahta, Textes et Etudes du Moyen Age, 35 (Louvain-la-neuve: Féderation Internationale des Instituts d'Études Médiévales, 2006), pp. 559–83 (pp. 561, 562).

6 Literature on *translatio* is extensive and a full bibliography is impracticable here. For a brief discussion and further references, see Natalia I. Petrovskaia, '*Translatio* and Translation: The Duality of the Concept from the Middle Ages to the Early Modern Period', *Doshisha Studies in English*, 99 (2018), 119–36, and Natalia I. Petrovskaia, 'Translation and Transmission of Texts in Medieval Europe: Two Aspects of Translatio' in *Literature, Science and Religion: Textual Transmission and Translation in Medieval and Early Modern Europe*, ed. by Manel Bellmunt Serrano, Joan Mahiques Climent, Problemata Literaria, 88 (Kassel: Edition Reichenberger, 2020), pp. 359–74.

7 See, for instance, A. G. Jongkees, 'Translatio Studii: Les avatars d'une thème médiéval', *Miscellanea Medievalia in Memoriam Jan Frederick Niermeyer* (Groningen: J. B. Wolters, 1967), pp. 41–51; S. Lusignan, 'La topique de la *translatio studii* et les traductions françaises de textes savants au XIVe siècle', in *Traduction et traducteurs au Moyen Âge: Actes du colloque international du CNRS organisé à Paris. Institut de recherche et d'histoire des textes les 26–28 mai 1986* (Paris, 1989), pp. 303–15, and further bibliographical references provided in the discussion below.

8 Quoted and translated in Simon Gaunt, *Marco Polo's 'Le Devisement du Monde': Narrative Voice, Language and Diversity* (Cambridge: D. S. Brewer, 2013), p. 71 n. 66. Gaunt quotes Prior's edition of the *Image du monde*.

9 See Claude Buridant, '*Translatio medievalis*: Théorie et pratique de la traduction médiévale', *Travaux de linguistiaue et de littéature*, 21 (1983), 83–136, and more recently Petrovskaia, 'Translation and Transmission'. For an alternative view, see, for instance, Giacinta Spinosa, '*Translatio studiorum* through Philosophical Terminology', in *Translatio studiorum: Ancient, Medieval and Modern Bearers of Intellectual History*, ed. by Marco Sgabi (Leiden and New York: Brill, 2012), pp. 73–89 (p. 82). Spinosa distinguishes linguistic translation from *translatio*.

of *translatio studiorum* on the one hand, and of its origin, the translation concept inherent in the idea *translatio imperii*, on the other.[10] Giacinta Spinosa observes that the notion of cultural transfer and appropriation was inherent in the concept even before the conventionally accepted date of the formulation of *translatio imperii* idea in the ninth century, referring to Boethius.[11] The transmission of ancient knowledge in this view, as summarised by Spinosa, 'meant preparing a series of introductions and abstracts, both summing up what had been said too elaborately, and elaborating what had been said too succinctly'.[12] This applies to most of the *Imago mundi* family.[13]

The *Imago mundi* tradition is largely typical of attitudes towards authority and transmission of information seen in other texts. For instance, all the rewritings of the text share a cumulative attitude to the original. This attitude we find articulated in the thirteenth century by John of Salisbury in relation to twelfth-century teachers whose method was to rephrase the material in simpler terms and to supplement it from other sources.[14] This family of texts also provides an excellent illustration of the developments in the organisation of knowledge identified for the twelfth and thirteenth centuries by Malcolm Parkes.[15] The manuscripts of the original Latin text tend to have rubrics in the margins, but this is not consistent across the entire manuscript tradition. This is put to good use in the parallel-text Italian translation, preserved in a single manuscript at the Estense library, which uses the rubrics to split the text into manageable chunks to organise its two parallel language versions.[16] Meanwhile, some of the later rewritings have

10 For the connection between the two, see, for instance Christian Bratu, '*Translatio*, autorité et affirmation de soi chez Gaimar, Wace et Benoît de Sainte-Maure', in *The Medieval Chronicle VIII*, ed. by Erik Kooper and Sjoerd Levelt (Leiden: Brill, 2013), pp. 135–64; Tullio Gregory, 'Translatio Studiorum', in *Translatio Studiorum: Ancient, Medieval and Modern Bearers of Intellectual History*, ed. by M. Sgarbi (Leiden: Brill, 2014), pp. 1–22; Spinosa, '*Translatio studiorum*', pp. 73–74. See also Jongkees, 'Translatio Studii'.

11 Spinosa, '*Translatio studiorum*', pp. 74–75.

12 Spinosa, '*Translatio studiorum*', p. 75.

13 See Rita Copeland's observations on Brunetto Latini's *Trésor*; Copeland, 'Language Frontiers, Literary Form, and the Encyclopedia', in *Frontiers in the Middle Ages*, ed. by Merisalo and Pahta, pp. 507–24 (p. 517).

14 Discussed in Karen Pratt, 'Medieval Attitudes to Translation and Adaptation: The Rhetorical Theory and the Poetic Practice', in *The Medieval Translator II*, ed. by Roger Ellis (London: Queen Mary University of London, 1991), pp. 1–27 (pp. 10–11).

15 Malcolm Parkes, 'The Influence of the Concepts of *Ordinatio* and *Compilatio* on the Development of the Book', in *Medieval Learning and Literature. Essays Presented to Richard William Hunt*, in J. J. G. Alexander and M. T. Gibson (Oxford: Clarendon Press, 1976), pp. 115–41.

16 See p. 65.

marked chapter divisions, collected into chapter lists (and reorganise the text somewhat).[17] This reflects the increased sophistication of thirteenth-century *compilatio*, as, for example, in Vincent de Beauvais, Bartholomaeus Anglicus, and Brunetto Latini.[18] The *Imago mundi* tradition grows and evolves together with the organisation and compilation techniques.

The role of *translatio studii* in the self-positioning of these texts should not be underestimated, even when it is not explicitly expressed. For instance, in the French rewriting, the *Image du monde*, the account of the transmission of *clergie* (clerical learning) to France is coupled with the self-identification of the work as a vernacular *livre de clergie*, positioning itself within the process of *translatio*.[19] This is perhaps the most illustrative example of the identity, of the concepts of translation and transmission in medieval discourse.[20] The medieval 'translator' transmits the text, intepreting it (whether as interlingual translation or as an explanation, or modernisation of the same language) for the new audience. In some cases this process transforms the translator into an author(ity).[21] In the *Imago mundi* tradition this happens in particular to some of its French adaptations, which in turn become progenitors of further translations and adaptations.[22] This can be linked to the process of the gradual acceptance of vernacular texts as authorities in the late Middle Ages in parallel to the established authority

17 The *Image du monde*, for instance. Compare the discussion in Parkes, 'The Influence', pp. 118, 121, 123.

18 Parkes, 'The Influence', pp. 129–30.

19 Petrovskaia, 'Mythologizing the Conceptual Landscape: Religion and History in *Imago Mundi, Image du Monde* and *Delw y Byd*', in *Landscape and Myth in North-Western Europe*, ed. by Matthias Egeler, Medieval Texts and Cultures of Northern Europe (Turnhout: Brepols, 2019), pp. 195–211 (pp. 206–07); Natalia I. Petrovskaia, 'Translation and Transmission', pp. 370, 384, with reference to Claire M. Waters, *Translating Clergie: Status, Education, and Salvation in Thirteenth-Century Vernacular Texts* (Philadelphia: University of Pennsylvania Press, 2016) and Chantal Connochie-Bourgne, 'Pourquoi et comment réécrire une encyclopédie? Deux rédactions de l'*Image du monde*', in *Encyclopédies médiévales: discours et savoirs*, ed. by B. Baillaud, J. de Gramont, and D. Hüe, Cahiers Diderot, 10 (Rennes: Presses Universitaires de Rennes et Association Diderot, 1999), pp. 143–54.

20 The metaphorical use of 'translation' in postmodern discourse to signify cultural exchange of various non-linguistic types is a counter-movement; Bassnett, 'From Cultural Turn to Translational Turn', p. 74.

21 Cicero's transformation from translator to authority in Jean of Antioche's translation of Cicero's *Rhetoric* is particularly illustrative; Caroline Boucher, 'De la *subtilité* en français: Vulgarisation et savoir dans les traductions d'*auctoritates* des XIIIe–XIVe siècles', in *The Theory and Practice of Translation in the Middle Ages*, ed. by R. Voaden, R. Tixier, T. Sanchez Roura, and J. R. Rytting (Turnhout: Brepols, 2003), pp. 89–99 (p. 92).

22 For discomfort of the medieval translator's position negotiating creativity and transmission see Machan, 'Chaucer'; Petrovskaia, 'Translation and Transmission', pp. 368, 371–72.

of Latin texts.[23] If we are to regard the intellectual encyclopedic culture of the western Middle Ages as extending beyond a single linguistic vernacular environment, interlingual translation becomes part of rewriting as an 'act of cultural memory'.

The Latin text was adapted, and the adaptations were in turn translated, sometimes in several stages. Numerous works of immense historical importance and acknowledged cultural significance have translated, quoted, and used parts of our Latin text, from the Welsh *Delw y Byd*, the *Mappemonde* of Pierre de Beauvais, and Jacob van Maerlant's Dutch masterpiece *Alexanders geesten*, through the spectacular fourteenth-century Catalan Atlas/Ecumene Chart produced by the Majorcan cartographer Elisha ben Abraham Cresques, through the mid-fifteenth-century Spanish masterwork *Laberinto de Foruna* of Juan de Mena, to William Caxton's *Mirror of the World*.[24]

Figure 1 provides a rough timeline. It only includes the texts further mentioned in this book. Because the exact dates are known for only some of these, the relative chronology is necessarily approximate. The purpose of the image is to give an indication of of the tradition's extent and the complexity of the interrelations therein. Some of these texts gave rise to further adaptations. In Figure 1, examples include the *Divisiones* of Perot de Garbelei, derived from the *Mappemonde* of Pierre de Beauvais, and the

23 As Alistair J. Minnis points out, 'it is inaccurate and misleading to think of vernacular texts as having displaced Latin ones, in respect of prestige – as if *auctoritas* was a finite commodity, whose increment in one area meant its diminuition in another'; *Translations of Authority in Medieval English Literature: Valuing the Vernacular* (Cambridge: Cambridge University Press, 2009), p. 11. For attitudes to authority in geographical contexts, see Aníbal A. Biglieri, *Las ideas geográficas y la imagen del mundo en la literatura española medieval*, Medievalia Hispanica, 17 (Madrid: Iberoamericana; Frankfurt am Main: Vervuert, 2012), pp. 32–33.

24 Gabriel Llompart i Moragues et al., eds. *El Atlas Catalan de Cresques Abraham* (Barcelona: Diafora, 1975); Gabriel Llompart i Moragues et al., eds., *El mon i els dies: L'Atles Català* (Barcelona: Enciclopedia Catalana, 2005). For editions of the *Laberinto*, see, for example, Juan de Mena, *Laberinto de Fortuna*, ed. by John G. Cummins (Madrid: Cátedra, 1979) and Juan de Mena, *Laberinto de Fortuna*, ed. by Maxim P. A. M. Kerkhof, Nueva Biblioteca de Erudición y Crítica, 9 (Madrid: Castalia, 1995). María Rosa Lida de Malkiel, *Juan de Mena, poeta del prerrenacimiento español* (Mexico: Publicaciones de la Nueva Revista de Filología Hispánica, 1950) remains a key discussion of Mena and his opus. See also Florence Street, 'La vida de Juan de Mena', *Bulletin Hispanique*, 55 (1953), 149–73; Dorothy Clotelle Clarke, *Juan de Mena's 'Laberinto de Fortuna': Classic Epic and Mester de Clerecía*, Romance Monographs, 5 (University, MS: University of Mississipi, 1973); Cristina Moya García, ed., *Juan de Mena: De letrado a poeta* (Cambridge: Boydell & Brewer, 2015). For a discussion of Mena's use of the *Imago mundi* see now also Natalia I. Petrovskaia, 'The Limits of Medieval Translation: Juan de Mena's Use of the *Imago mundi* in the *Laberinto de Fortuna*', *Translation Matters*, 5 (2023), 19–30, DOI: 10.21747/21844585/tm5_2a2 (accessed 4 July 2024). William Caxton, *Mirrour of the World*, ed. by Oliver Herbert Prior, EETS, e.s. 110 (London, 1913); N. F. Blake, *William Caxton and English Literary Culture* (London: Hambledon, 1991).

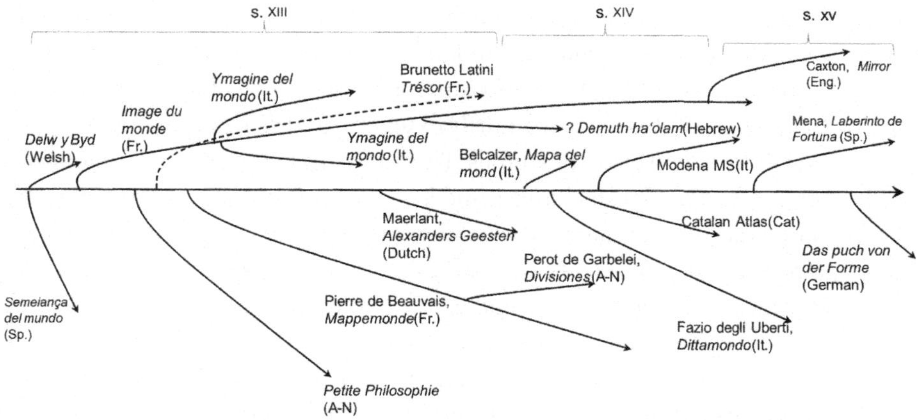

Figure 1: Timeline of *Imago mundi* adaptations

Italian, Middle English, and Hebrew adaptations derived from the *Image du monde* by Gautier or Gossouin de Metz. For some authors it is not entirely clear whether they used the *Imago mundi* or an adaptation thereof. These texts are introduced in greater detail in the following sections of this chapter. In Latini's case I have drawn a dotted line from the Latin crossing the line of the French because his source is unclear.

Several branches of the tradition examined here were enriched with a (sometimes quite complex) programme of illustrations. The *Image du monde* was translated and transmitted into Middle English alongside with its programme of illustrations by William Caxton (?1420s–*c.* 1492) at the end of the fifteenth century (1481, second printing *c.* 1490).[25] The *mapa del mond* of Vivaldo Belcalzèr (*c.* 1250–*c.* 1312), incorporated into his translation of Bartholomaeus Anglicus's *De proprietatibus rerum*, is also preceeded by a *mappa mundi*.[26] As Baxter points out, medieval *mappae mundi* also often bear quotations from the *Imago mundi*.[27] The most spectacular example is the atlas in Paris, Bibliothèque nationale de France (BnF), MS Esp. 30 (s. XIV), known as the Catalan Atlas (now also known as the Ecumene Chart).[28]

In addition to the texts presented in Figure 1, multiple authors used the text at least marginally in their work. Indeed, it has been tentatively

25 Blake, *William Caxton and English Literary Culture*, esp. pp. 1–36 for Caxton's biography and pp. 103–04, 108–11 for the *Mirror of the World*; Blake, *Caxton: England's First Publisher* (London: Barnes and Noble, 1976), pp. 135–38 for *Mirror of the World*.

26 See p. 73 n. 87.

27 Jason Baxter, '"Videmus nunc per speculum": The Mysticism and Naturalism of the Twelfth-Century *imago mundi*', *Haskins Society Journal*, 28 (2017), 119–42 (p. 128).

28 See p. 74 of this book.

suggested that traces of the *Imago mundi* might be seen even in Dante Alighieri's *Divina Commedia* and that Honorius might be placed among the inspirations of Christopher Columbus.[29] We will be able to discuss only a small part of this extraordinarily extensive and rich corpus. The rest must await its own (series) of studies. The rewritings described in the following chapters make both changes and additions to the original text, and their new status as new authorities underlines the importance of these changes. The range of approaches is striking. While some, like *Delw y Byd*, translate faithfully, the dominant trend is to expand the text and appropriate it to its new cultural and linguistic (literary) context. Some, like the *Semeiança del mundo*, make minimal alterations, such as introduction of transitional passages between sections of text or addresses to the audience, which bring the resulting text in line with the host literary culture. Others, such as Gossouin's *Image du monde*, use *Imago mundi* I as the centre stone of a new encyclopedic work.[30] As Rita Copeland observes, the combination of narrative (Copeland refers specifically to 'life-narrative') and encyclopedic material is a way for vernacular writers to 'respond to the encyclopedia as literary form'.[31]

The discussion in this chapter aims to provide an overview of the types of responses elicited by the *Imago mundi* in the various rewritings. For this purpose, I introduce four categories:

(1) Vernacular translations proper;
(2) Vernacular adaptations in encyclopedic contexts (this includes faithful translations of a section of the text where they are incorporated into new and broader encyclopedic texts);

29 Marie-Odile Garrigues, 'Honorius Augustodunensis et l'Italie', *Mélanges de l'École Française de Rome. Moyen Âge, Temps Modernes*, 84 (1952), 511–30 (pp. 524–25, 527–29). For Pierre d'Ailly's *Image du monde* and Columbus, see Valerie I. J. Flint, *The Imaginative Landscape of Christopher Columbus* (Princeton, NJ: Princeton University Press, 1992), pp. 44–49, 54, 129, 140, 168; and the early study by George E. Nunn, 'The *Imago mundi* and Columbus', *The American Historical Review*, 40 (1935), 646–61. See also Theodore J. Cachey Jr., 'Cosmology, Geography, and Cartography', in *Dante in Context*, ed. by Zygmunt G. Barański and Lino Pertile (Cambridge: Cambridge University Press, 2015), pp. 221–40 (pp. 227, 230, 235); and Luis M. Girón Negrón, 'Islamic and Jewish Influences', in *Dante in Context*, ed. by Barański and Pertile, pp. 200–20 (pp. 204, 212).

30 See above, pp. 55, 57, 59.

31 Copeland's example Latini's *Tesoretto*, which uses the story of the circumstances of its own composition; 'Language Frontiers', pp. 517–18. For the text see Latini, *Il tesoretto: The Little Treasure*, ed. and trans. by J. B. Holloway, Garland Library of Medieval Literature, 2 (New York: Garland, 1981).

(3) Vernacular adaptations in non-encyclopedic contexts (this includes faithful translations of a section of *Imago mundi* where they are incorporated into texts that belong to other genres, such as romance);
(4) Adaptations into new Latin contexts.

Inevitably, some overlap will be seen between these categories. In that respect it is useful to think of these, too, as fuzzy sets. Each category is described in a separate subsection below for ease of reference.

Figure 2 provides a visual representation of the categories as fuzzy sets using an opacity-varying diagram modelled in principle on a Venn diagram.[32] It shows overlap between categories of vernacular translations/adaptations and none between vernacular and Latin categories. One could represent the latter distinction with a non-transparent linear border, since the difference in languages provides a clear boundary between these groups, but I have preferred not to do this here because even though these particular fuzzy sets do not overlap with each other, they can overlap with other fuzzy sets (other Latin encyclopedias, for instance).

Since this is the first study bringing together the enormous *Imago mundi* textual family into a single discussion, it has been necessary to be selective in the choice of texts for detailed analysis due to the size of the tradition, the lack of up-to-date editions or detailed studies of a number of the texts involved, and limitations of space.[33] The selection of texts for in-depth

32 This diagram is not mathematically generated, as that level of precision is not required for the illustrative purposes it fulfils. However, it is possible to generate such a diagram mathematically for fuzzy sets; see Lifeng Zhu, Weiwei Xia, Jia Liu, and Aiguo Song, 'Visualizing Fuzzy Sets Using Opacity-Varying Freeform Diagrams', *Information Visualization*, 17 (2017), 146–60.

33 Valerie I. J. Flint, 'Honorius Augustodunensis', in *Authors of the Middle Ages II: Historical and Religious Writers of the Latin West* Nos 5–6, ed. by Patrick J. Geary (Aldershot: Variorum Reprints, 1995), pp. 89–183 (pp. 165–67); Michael W. Twomey, 'Medieval Encyclopedias', in R. E. Kaske, *Medieval Christian Literary Imagery. A Guide to Interpretation* (Toronto: University of Toronto Press, 1988), pp. 182–215 (pp. 189–91). The German translation of the *Imago mundi*, for instance, which to my knowledge is preserved in only one manuscript, Berlin, Staatsbibliothek Preussischer Kulturbesitz, Germ. Fol. 595 (s. XIV), still awaits its editor. The reference was found by Kiki Calis. For more details, see entries in the *Images of the World* database, https://imagomundi.hum.uu.nl/ (accessed 28 March 2020); the manuscript is described in Hermann Degering, *Kurzes Verzeichnis der germanischen Handschriften der Preussischen Staatsbibliothek* I *Die handschriften in Folioformat* (Leipzig: Verlag von Karl W. Hiersemann, 1925), p. 66. Notable absentees from this list include the *Liber de existencia* (c. 1200), which relies heavily on *Imago mundi*, but combines it with other sources in a complex pattern (borrowing individual phrases from different texts and re-combining them). For a discussion, see Patrick Gautier Dalché, ed., *Carte marine et portulan au XIIe siècle: Le Liber de Existencia Riverierarum et Forma Maris Nostri Mediterranei (Pise, circa 1200)* (Rome: l'École Française de Rome, 1995), pp. 89–90, 104; his

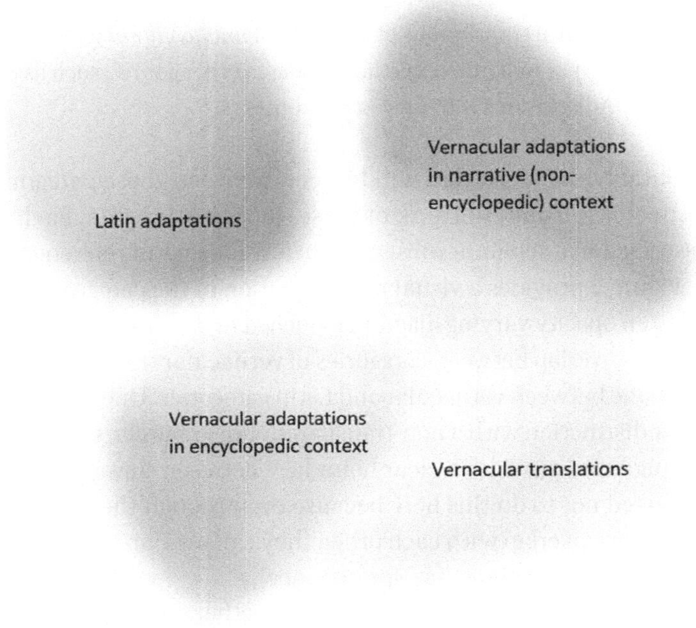

Figure 2: Categories of *Imago mundi* adaptations visualised in an opacity-varying diagram

discussion has been guided by considerations of geographical scope and generic (in the modern understanding of genre) range.

One of the objectives here is to provide a sense of the range and diversity of the texts which contain geographical description of Europe translated or derived from the *Imago mundi*, and to elucidate, insofar as it is possible in the course of a necessarily cursory examination, the relationships between some of these. This study is intended as a prolegomenon to further investigations.

Vernacular Translations of *Imago mundi*

The thirteenth century was a high point for the production of new *Imago mundi* rewritings. Two faithful translations (in the modern sense of 'translation') of Book I were produced in thirteenth-century Spain and Wales, and in Germany possibly a century later. An Italian translation also survives in a fourteenth-century manuscript, where it accompanies the Latin text as

edition annotates the sources, identifying sections of Honorius's *Imago mundi* used throughout the text.

alternating paragraphs.[34] A further set of faithful translations of this text were produced in this period, but those that were incorporated into other texts are discussed separately, in the following sections. Here I introduce only those translations which retain the unity of the original text.

Three Iberian adaptations were produced in the Middle Ages. Two are represented by translations of the *Imago mundi* incorporated into new contexts and are classified with non-encyclopedic texts below.[35] The translation proper is *Semeiança del mundo* (1173×1223).[36] The title is modern, as the manuscripts entitle the text *Mapa mundi* and attribute it to Isidore.[37] Although the *Semeiança* survives in four manuscripts (three medieval), there are at least two versions, which despite minor differences probably represent recensions of a single translation.[38] It is worth noting that despite its misattribution to Isidore, in one of the three medieval manuscripts the *Semeiança* is preceded by a Castilian translation of Honorius's *Elucidarium*.[39] While it does not necessarily indicate these texts were thought to have common authorship, it implies that they were read together and seen, at

34 The manuscript is Biblioteca Estense MS α.Q.5.1. The translation and its function are discussed in Petrovskaia, 'Explicit and Implicit Multilingualisms: *Imago mundi* and MS Estense α.Q.5.1', in *Multilingual Dynamics of Medieval Literature in Western Europe, c. 1200–c. 1600*, ed. by Bart Besamusca, David Murray, and Lisa Demets (Turnhout: Brepols, forthcoming).

35 Juan de Mena's *Laberinto* and Jehuda ben Abraham Cresques's *Catalan Atlas*.

36 William E. Bull and Harry F. Williams, eds, *Semeiança del mundo: A Medieval Description of the World* (Berkeley and Los Angeles: University of California Press, 1959), p. 1. A close comparison of the text with the *Imago mundi, pace* the text's editors, who state that it is based on Isidore's *Etymologies*, yields the conclusion that it is largely a verbatim translation of Honorius's text. For a recent discussion of the qualities of this text as translation, see Marta Lacomba, 'La articulación de un discurso científico en castellano bajo Alfonso X. Se l *Semeiança del mundo* al *Lapidatio*: De una descripción del mundo en romance a la construcción de un espacio castellano del saber' in *Lenguas, reinos y dialectos en la Edad Media Ibérica: La construcción de la identidad: homenaje a Juán Ramón Lodares* (Madrid: Iberoamericana; Frankfurt am Main: Vervuers, 2008), pp. 341–66, esp. pp. 342–43, 351, 356–59.

37 Bull and Williams, eds, *Semeiança del mundo*, p. 12.

38 Discussed in Bull and Williams, eds, *Semeiança del mundo*; Richard P. Kinkade, 'Un nuevo manuscrito de la *Semeiança del mundo*', *Hispanic Review*, 39 (1971), 261–70; Lacomba, 'La articulación de un discurso científico', p. 343; see also below, pp. 64, 98–99, 121–25, 132.

39 Richard Kinkade, ed., *Los Lucidarios españoles: Estudio y edición* (Madrid: Gredos, 1968). Madrid, Biblioteca Nacional MS 3369 (s. XIV–XV) is one of six *Lucidario* manuscripts; Kinkade, 'Un nuevo manuscrito'; Luca Sacchi, 'L'edizione di testi modulari: Il caso del *Lucidario* di Sancho IV di Castiglia', in *Prassi ecdotiche: Esperienze editoriali su testi manoscritti e testi a stampa*, ed. by A. Cadioli and Paolo Chiesa (Milan: Cisalpino, 2008), pp. 155–70. The two other *Semeiança* manuscripts, Lisbon, Biblioteca Nacional, MS 46 (s. XIII) and Madrid, Biblioteca del Escorial X.III.4 (Spain, 1467), are edited in Bull and Williams, *Semeiança del mundo*; see also Kinkade, 'Un nuevo manuscrito'.

least thematically, as connected.[40] Richard Kinkade has also suggested that the combination of two of Honorius's works in a single manuscript may indicate that a Latin exemplar containing a collection of Honorius's works may have served as base for the translation.[41]

While misattribution to eminent authorities is a common feature of medieval textual transmission, and fairly common for *Imago mundi* and related texts, the choice of Isidore in Spain might indicate appropriation.[42] This type of appropriation is visible elsewhere in the *Imago mundi* tradition. Multiple manuscripts of English provenance attribute the text to Anselm, Henry of Huntingdon, and Bede.[43] Although this suggestion remains tentative, pending a full survey mapping (mis-)attributions onto manuscript provenance, this might represent a tendency for appropriation of the geographical framework provided by the text through its re-attribution to a local authority. In the case of Spain, this re-attribution to Isidore ultimately reinforces the text's actual genealogy, since, as mentioned above, Isidore's encyclopedia forms the backbone of Honorius's.[44]

Kinkade notes that the variation between the three surviving manuscripts of the *Semeiança* makes it difficult to establish an Ur-text. In the context of other translations and adaptations of the *Imago mundi* examined here it is worth raising the possibility that these surviving manuscripts may represent more than one translation, or at the very least involve interpolations based on a redactor consulting the Latin material. The situation is not unparalleled, witness the two independently made Welsh translations and the multiple Anglo-Norman and French rewritings of Honorius's text.

An additional difficulty in analysing the cultural and intellectual context of the *Semeiança* is the result of the currently accepted dating of the text. If the translation was indeed completed in 1223, it pre-dates the great floruit of scientific text production and translation activity that took place later in the century under the patronage of Alfonso X 'el Sabio'.[45] If the different

40 See also Kinkade's suggestion that the texts were translated from a collection of Honorius's works; Kinkade, 'Un nuevo manuscrito', p. 268 n. 9.

41 Kinkade, 'Un nuevo manuscrito', p. 268 n. 9.

42 Note, however, that the choice of Isidore may also have been dictated simply by virtue of the latter's status as the ultimate encyclopedic authority; cf. Lacomba, 'La articulación de un discurso científico', p. 343.

43 Authorship attributions can be searched for in the *Images of the World: Manuscripts Database of the Imago Mundi Tradition* https://imagomundi.hum.uu.nl/ (accessed 4 July 2024).

44 See above, pp. 40–41, 45. Flint identifies Bede's *De natura rerum* as one of Honorius's sources.

45 The literature on this subject is vast. A good place to start for discussions and a further bibliography is Robert I. Burns, ed., *Emperor of Culture: Alfonso X the Learned of Castile and His Thirteenth-Century Renaissance* (Philadephia: University of Pennsylvania Press, 2015)

manuscripts represent different versions, however, it might be worth asking whether any of these reworkings were undertaken during Alfonso's reign.

In Wales also at least two separate translations of *Imago mundi* I were produced sometime in the first half of the thirteenth century and are now known as *Delw y Byd*.[46] No manuscript contains a Welsh translation of the whole of Book I. However, although the text survives only in an incomplete form, it is worth noting that the comparatively small number of medieval manuscripts (six) that preserve it is large by Welsh standards. Together with the fact that these six manuscripts carry two separate translations, the number of witnesses attests to the text's popularity in Wales.[47] The translations demonstrate only very minor alterations to the description of Europe, particularly in the updating of a number of place-names which may have been of local interest.[48] They thus have a minor role to play in *Transforming Europe*.

Two further faithful translations of the text were produced at a later period, one in Italy and one in Germany. The Italian translation preserved in Modena, Biblioteca Estense, MS α.Q.5.1 (s. XIV), is in the Venetian dialect and uniquely presents a parallel text with a vernacular translation following each Latin chapter.[49] This is one of several Italian translations and adaptations of the text.[50] The German translation of *Imago mundi* I, which survives in a single fifteenth-century manuscript, where it is entitled *Das puch von der Forme und pildnuss der Welt* ('The Book of the Form and Image of the World'), remains unedited and unstudied.[51] In the manuscript,

46 See Petrovskaia, ed., *Delw y Byd*, p. 7; and for general overview of the tradition, pp. 6–10, 14–26.

47 There is also at least one manuscript of *Imago mundi* of Welsh provenance surviving, the *Imago mundi* produced in Wales, Exeter, Cathedral Library MS 3514 (s. XIII); for discussion, Julia Crick, 'The Power and the Glory: Conquest and Cosmology in Edwardian Wales (Exeter, Cathedral Library, 3514', in *Textual Cultures, Cultural Texts*, ed. Orietta da Rold and Elaine Treharne (Cambridge: Brewer, 2010), pp. 21–42, and Natalia I. Petrovskaia, *Medieval Welsh Perceptions of the Orient* (Turnhout: Brepols, 2015), pp. 8, 11.

48 See discussion in Natalia I. Petrovskaia, 'The Concept of Europe in the Medieval Welsh Geographical Treatise *Delw y Byd*', *Celtic Forum*, 21 (2018), 23–34.

49 Its text was edited in Vittorio Finzi, 'Di un inedito volgarizzamento dell'*Imago mundi* di Onorio d'Autun', *Zeitschrift für Romanische Philologie*, 17 (1893), 490–543 and 18 (1894), 1–73.

50 For the others, see below, p. 70.

51 In the manuscript it follows upon a German chronicle, which dates the manuscript to around 1480/1490, and is bound together with part of an earlier manuscript, containing the Latin text of the *Imago mundi* (incomplete, breaking off partway through Book III); Peter Jörg Becker and Eef Overgaauw, eds, *Aderlass und Seelentorst: Die Überlieferung deutscher Texte im Spiegel Berliner Handschriften und Inkunabeln* (Mainz am Rhein: Verlag Philipp von Zabern for Staatsbibliothek zu Berlin, 2003), pp. 429–31.

Berlin, Staatsbibliothek Preussischer Kulturbesitz, Germ. Fol. 595, it follows upon a chronicle which ends at 1471, thus setting a *post quem* date for the copy. The date of the translation itself, however, is undetermined and must await detailed linguistic analysis.[52] A cursory examination shows that this text appears to follow the Latin text very closely, without any additions, omissions, or alterations.

There is some evidence that an Old Norse translation of the *Imago mundi* may have existed at some point, but only fragments survive.[53] The surviving fragment is a loose translation of *Imago mundi* I.1, presenting the metaphor of the world as an egg, with some additional information on solar and lunar rotation, and is probably a copy of an earlier, fourteenth-century translation.[54] It survives in several manuscripts: Árni Magnússon Institute AM 746 4to, ff. 2r–v (s. XIV), AM 685d 4to, f. 31r (s. XV), with the fullest text in AM 238 fol. XIX, ff. 1r–v (s. XV).[55] AM 685d refers explicitly to the text by title (*svá segir Imago mundi at heimrinn sé vaxinn sem egg*).[56] In addition, the manuscript Copenhagen, Det Kongelige Bibliothek, Gml. kgl. Sml. 1812 4to, at ff. 2v and 10v–11r, appears to be a vernacular summary of *Imago mundi* I.74–81.[57] Although none of these extracts contain references to Europe which would interest us, they demonstrate the presence of *Imago mundi* in the Old Norse cultural sphere, and provide context for

52 For a description of the MS and further bibliography, see also https://handschriftenportal.de/search?hspobjectid=HSP-58604a40-ca06-3be3-ba16-1818be1bfe7b (accessed 4 July 2024). Schorbach describes it as a fifteenth-century prose translation, based on the date of the manuscript (citing the library only, but presumably the reference is to our MS); *Studien*, p. 163.

53 Margaret Clunies Ross and Rudolf Simek, 'Encyclopedic Literature', in *Medieval Scandinavia: An Encyclopedia*, ed. by P. Pulsiano, Kirsten Wolf, Paul Acker, and Donald K. Fry (New York and London: Garland, 1993), pp. 164–66 (p. 166). This should be taken in the context of the fragmentary nature of much of the manuscript tradition in the region; for a general discussion, see Åslaug Ommundsen and Tuomas Heikkilä, eds, *Nordic Latin Manuscript Fragments. The Destruction and Reconstruction of Medieval Books* (London and New York: Routledge, 2017).

54 Clunies Ross and Simek, 'Encyclopedic Literature', p. 166.

55 For the texts, see Rudolf Simek, *Altnordische Kosmographie: Studien und Quellen zu Weltbild und Weltbeschreibung in Norwegen und Island vom 12. bis zum 14. Jahrhundert* (Berlin and Boston: De Gruyter, 2012), pp. 396–98. See the discussion in Elise Kleivane, 'There is More to Stjórn than Biblical Translation', in *Speculum septentrionale: Konungs skuggsjá and the European Encyclopedia of the Middle Ages*, ed. by Karl G. Johansson and Elise Kleivane (Oslo: Novus Forlag, 2018), pp. 115–47.

56 *Alfræði Íslenzk*, ed. by K. Kålund and N. Beckman, 3 vols. (København: Samfund til Udgivelse af gammel nordisk Litteratur, 1908–18), III, p. 75.

57 Simek, *Altnordische Kosmographie*, p. 400; Honorius Augustodunensis, *Imago mundi*, ed. by Valerie I. J. Flint, *Archives d'histoire doctrinale et littéraire du Moyen Âge*, 49 (1982), pp. 76–78. Simek corrects the number of years Mars needs to complete its course from two to twelve, but it is the erroneous 'two' in the *Imago mundi*; cf. Flint, ed., *Imago mundi*, p. 78.

its traces in the vernacular literature, which speak of the place of Europe and of Scandinavia within the world.[58] We return to these in Chapter 3. Here, for the sake of context I must note that various works by Honorius had been translated, in full or in part, into Old Norse and the information contained therein was referred to in various Old Norse works, both of literary and technical nature.[59] Further, as Rudolf Simek observes, the *Imago mundi* was one of several important Latin encyclopedias known in medieval Iceland.[60]

Adaptations of *Imago mundi* I and Insertion of Translated Extracts in Encyclopedic Contexts

The translations discussed above, faithful to the original, were produced across Europe in various vernaculars throughout the period covered in this book. These are far outnumbered, however, by loose adaptations, or, to use Plate and Rose's term, rewritings. Francophone rewritings of the *Imago mundi* are numerous, though none can be termed translations.

The first of the French versions was composed towards the end of the twelfth or the beginning of the thirteenth centuries by Pierre de Beauvais under the title *Mappemonde*, rendering *Imago mundi* I.[61] A second French

58 For an example illustrating the presence of the type of worldview represented by the *Imago mundi* tradition in Old Norse, see Matthias Egeler, 'Eikþyrnir and the Rivers of Paradise: Cosmological Perspectives on dating *Grímnismál* 26–28', *Arkiv för nordisk filologi*, 128 (2013), 17–39. Paul Lehmann notes that many encyclopedic, cosmographic, and geographic texts which were popular in Iceland no longer survive in Icelandic manuscripts, and are attested only as citations, quotations, and extracts; Lehmann, 'Skandinaviens Anteil an der Lateinischen Literatur und Wissenschaft des Mittelalters', in Paul Lehmann, *Erforschung des Mittelalters: Ausgewählte Abhandlungen und Ausätze*, 5 vols (Stuttgart: Anton Hiersemann, 1959–73), vol. 5 (1962), pp. 275–393 (p. 354).

59 A bibliography of discussions of the knowledge of work by Honorius in medieval Iceland is provided in David McDougall, 'A Note on the Shape of Rome in the *Fóstbrœðra saga*', *Gripla*, 6 (1984), 259–64 (p. 264 n. 12).

60 Simek, *Altnordische Kosmographie*, p. 26. See also Paul Lehman, 'Skandinaviens Anteil', pp. 343, 358.

61 Annie Angremy, 'La *Mappemonde* de Pierre de Beauvais', *Romania*, 104 (1983), 316–50 (p. 319); see also George L. Hamilton, 'Encore un plagiat médiéval: La *Mappemonde* de Pierre de Beauvais et les *Divisiones mundi* de Perot de Garbelai', in *Mélanges de linguistique et de littérature offerts à M. Alfred Jeanroy par ses élèves et ses amis* (Paris: Droz, 1928), pp. 627–38, and the brief reference in Adrian Armstrong and Sarah Kay, with participation of Rebecca Dixon, Miranda Griffin, Sylvia Huot, Francesca Nicholson, and Finn Sinclair, *Knowing Poetry: Verse in Medieval France from the 'Rose' to the 'Rhétoriques'* (Ithaca, NY and London: Cornell University Press, 2011), p. 104.

version was produced by Gossouin (sometimes called Gautier) de Metz, in the mid-thirteenth century under the title *Image du monde*, which in turn survives in multiple versions and will be discussed further below, because it gave rise to its own set of rewritings. There are also two Anglo-Norman versions. An Anglo-Norman French version, *La Petite Philosophie*, was produced *c.* 1230.[62] On the basis of this, in the fourteenth century Perot of Garbelei (also spelled *Garbelai*) produced the *Divisiones mundi*, also in Anglo-Norman French.[63]

The *Image du monde*, the first version of which was produced *c.* 1245, is one of the most important members of the *Imago mundi* family, because it generated a series of translations and adaptations of its own.[64] This text incorporates *Imago mundi* I into a new construction, in the *livre de clergie* genre.[65] It has been generally accepted that the *Image du monde* was intended for a non-scholarly lay audience and as a simplified version of the original

62 Trethewey, ed., *La Petite Philosophie*; Twomey, 'Medieval Encyclopedias', p. 190; Armstrong et al., *Knowing Poetry*, p. 104.

63 O. H. Prior, ed., *Divisiones Mundi*, in *Cambridge Anglo-Norman Texts* (Cambridge: Cambridge University Press, 1924), pp. 33–62; Twomey, 'Medieval Encyclopedias', p. 190; Flint, ed., *Imago mundi*, pp. 9–10. Although this text may have used the Latin in part, it appears to have largely copied the *Mappemonde*; Hamilton, 'Encore un plagiat médiéval'; Angremy, 'La *Mappemonde*', pp. 331–35. Such cross-contamination highlights the complexities of the family history of *Imago mundi*–based texts.

64 For editions see Gossouin de Metz, *L'Image du monde, une encyclopédie du XIIIe siècle: edition critique et commentaire de la première version*, ed. by Chantal Connochie-Bourgne (unpublished PhD thesis, University of Paris, Sorbonne, Paris IV, 1999) (verse), and O. H. Prior, ed., *L'Image du monde de maitre Gossuin: Rédaction en prose* (Lausanne and Paris, 1913) (prose). For discussions, see, for instance, Connochie-Bourgne, 'Le cas de l'*Image du monde*: une encyclopédie du XIIIe siècle, ses sources antiques, l'apport médiéval', in *La transmission des connaissances techniques: tables rondes Aix-en-Provence, avril 1993–mai 1994*, ed. by M.-Cl. Amouretti and G. Comet, Cahier d'histoire des techniques, 3 (Aix-en-Provence: Publications de l'Université de Provence, 1995), pp. 87–98; Connochie-Bourgne, 'Pourquoi et comment réécrire une encyclopédie', pp. 143–54; Katherine A. Brown, 'The Vernacular Universe: Gossuin de Metz's *Image du monde*, *translatio studii*, and Vernacular Narrative', *Viator*, 44 (2013), 137–58. See also Sara Centili, 'La tradition manuscrite de l'*Image du monde*: fortune et diffusion d'une encyclopédie du XIIIe siècle' (unpublished PhD dissertation, Paris, Ecole des Chartes, 2005). I am grateful to Sara Centili for generously sharing with me chapters of her thesis.

65 Robémont, 'Statut de l'astronomie et evolution des connaissances sur le cosmos chez les vulgarizateurs médiévaux: le cas de quelques encyclopédies en langue vernaculaire', in *Observer, lire, écrire le ciel au Moyen Âge. Actes du colloque d'Orléans, 22–23 avril 1989*, ed. by Bernard Ribémont (Paris, 1991), pp. 283–300 (p. 291); Sara Centili, 'La seconda redazione in versi dell'*Image du monde*: una riscrittura didactica', *Cultura neolatina*, 66 (2006), 161–206 (pp. 184–85). For a discussion of the *Image du monde* as a *livre de clergie*, see Cenili, 'La seconda redazione', p. 163; Connochie-Bourgne, 'Images de la terre'; Connochie-Bourgne, 'Ordoner les éléments du savoir'; Petrovskaia, 'Mythologizing the Conceptual Landscape', p. 206.

material.[66] It makes explicit reference to the use of illustrations in making difficult material accessible to the general audience.[67] The text was enormously popular and continued to be copied well into the fifteenth century (subsequently making its way into print).[68]

The new text is also in three books, but adds an overview of the structure of knowledge and its transfer to France (invoking, but not naming the *translatio studii*) in the first book.[69] Gossouin used *Imago mundi* I as the main source of his Book II, but with some significant alterations, particularly in relation to the description of Europe, discussed in Chapter 4.[70]

The Europe section is comparatively short, comprising of 26 lines only, 'car souvent en oions parler' ('because we often hear of it spoken').[71] Note, however, that the subsequent section on Africa is equally short.[72] The description lists countries which compose Europe, with little additional information. No historical or biblical names are given, and only two non-geographical elements are specified.[73]

The text was dedicated to Robert d'Artois (Count Robert I of Artois), brother of Louis IX of France.[74] There are three or four versions of the *Image*

66 See, for instance, Brown, 'The Vernacular Universe', p. 138; Benoît Beyer de Ryke, 'Le miroir du monde: un parcours dans l'encyclopédisme médiéval', *Revue belge de philologie et d'histoire*, 81 (2003), 1243–74 (pp. 1266–67); Connochie-Bourgne, 'Pourquoi et comment réécrire une encyclopédie', p. 152; Petrovskaia, 'Mythologizing the Conceptual Landscape', p. 206.

67 Paris, BnF, MS Français 14961 (Limousin, s. XIII) is a good example of the typical layout and illustration pattern of the *Image du monde*; see also *Gallica* https://gallica.bnf.fr/ark:/12148/btv1b9061303j/f78.image (accessed 11 April 2020).

68 Baltimore, Walters Museum of Art MS W.199 (Bruges, 1489), roughly contemporary to Caxton's Middle English printed translation, on which more below, is an example of this. For the manuscript, see *The Digital Walters* http://www.thedigitalwalters.org/Data/WaltersManuscripts/html/W199/description.html (accessed 11 April 2024). For references to the print versions, see below, pp. 168–70.

69 There are three or four distinct versions of the *Image du monde*, of which one is in prose. The number depends on whether the text of London, British Library (BL), MS Harley 4333 (s. XIII) is recognised as a separate version. See Connochie-Bourgne, ed., *Image du monde*, pp. 38–43. For more on this manuscript, see also Paul Meyer, '*L'Image du monde*, rédaction du MS. Harley 4333', *Romania*, 21 (1892), 481–505.

70 Like the *Imago mundi*, the *Image du monde* is composed of three books, the second of which is so close to the Latin text as to be describable as, to a large extent, a translation of the other. Another alteration, the relabelling of the whole of Asia as *India*, is examined in Petrovskaia, 'Mythologizing the Conceptual Landscape'.

71 Connochie-Bourgne, ed., *Image du monde*, p. 839, l. 2980.

72 Also twenty-six lines; see discussion below.

73 A fountain which cannot extinguish flames, and a stone that cannot be extinguished when on fire (neither is given a religious explanation); *Image du monde*, ed. Connochie-Bourgne, pp. 839–40, ll. 2986–91.

74 See comment in Beyer de Ryke, 'Le miroir du monde, p. 1267 n. 91.

du monde, of which one is prose and the rest are verse.[75] Unlike those of *Imago mundi*, the manuscripts of *Image du monde* are often illuminated, and the illustrations form part of the original design of the work. There are over a hundred manuscripts of this text surviving, most of them illustrated.[76] The prose version appears to have been somewhat less widely circulated than the verse, with only nine manuscripts surviving.[77] However, of these one, a manuscript produced in Bruges in 1464 (BL MS Royal 19A IX) was used by William Caxton, who translated the text into Middle English as *Mirrour of the World* in 1481, the first English printed book with printed illustrations.[78] A second edition was published *c.* 1490.[79] Arguably dependent on the *Image du monde*, too, were several of the fourteenth-century Italian texts known as *L'Ymagine del mondo*, as well as a medieval Hebrew version, which has remained, however, little studied.[80] A further, apparently

75 Armstrong et al., *Knowing Poetry*, p. 104, suggest the second verse version post-dates the prose; cf. Connochie-Bourgne, ed., *Image du monde*, pp. 32–58. For attribution of all versions to Gossouin, see Christine Silvi, 'Les variations sur le nom dans la *Mappemonde* de maître Gossuin de Metz', in *Par les mots et les textes. Mélanges de langue, de littérature et d'histoire des sciences médiévales offerts à Claude Thomasset*, ed. by Danielle Jacquart, Danièle James-Raoul, and Olivier Soutet (Paris, 2005), pp. 679–91 (p. 691).

76 For a full survey of the manuscript tradition see Centili, 'La tradition manuscrite'.

77 The prose version is not identical to the preceding verse version in terms of contents. Some regional names are reordered between the second verse and the subsequent prose version of this text; Slivi, 'Les variations', pp. 681–82. One of these, in the conflation of Albania and Armenia in the prose version, is the result of a misreading by the compiler, according to Silvi (ibid., p. 682) with reference to Connochie-Bourgne, ed., *Image du monde*, l. 2763, and Prior, ed., *Image du monde*, p. 124.

78 The illustration scheme copies that of the French text; James A. Knapp, 'Translating for Print: Continuity and Change in Caxton's *Mirrour of the World*', *Disputatio*, 3 (1998): 64–90 (p. 66). For the edition, see the Library of Congress website: http://hdl.loc.gov/loc.rbc/rosenwald.0563.1 (accessed 12 April 2024). This text should not be confused with the similarly titled moral treatise in Oxford, Bodleian Library, MS Bodley 283 (s. xv); Robert R. Raymo and Elaine E. Whitaker, ed., *The Mirroure of the Worlde: A Middle English Translation of Le miroir du monde*, Medieval Academy Books, 106 (Toronto: University of Toronto Press for the Medieval Academy of America, 2003). For a general discussion and further bibliography on Caxton, see Blake, *Caxton*, and Lotte Hellinga, *William Caxton and Early Printing in England* (London: BL, 2010).

79 Both are available on *EEBO*. See https://www.proquest.com/books/hier-begynneth-book-callid-myrrour-worlde/docview/2264211121/se-2 (accessed 3 April 2023) and https://www.proquest.com/books/hier-begynneth-booke-callyd-myrrour-worlde/docview/2264193458/se-2 (accessed 3 April 2023).

80 For the Italian tradition see Finzi, ed., 'Di un inedito volgarizzamento', and Francesco Chiovaro, ed., *L'Ymagine del mondo (Firenze, Bibl. naz. cod. palat. 703)*, Quaderni partenopei, 1 (Naples: Loffredo, 1977). See also Twomey, 'Medieval Encyclopedias', p. 190. For the Hebrew text, see A. Neubauer, 'Les traductions hébraïques de l'*Image du Monde*', *Romania*, 5 (1876), 129–39. Note that the Hebrew translation of the *Image du monde* is not included in Mauro Zonta's

independent, translation into Hebrew, צל העולם (*Ẓel ha-'Olam / Ṣel ha-'olam*) 'Shadow of the World', was made in the early modern period by Matthias Delacrut, and attributed to the thirteenth century Hagin Deulacres.[81] It was first published in Amsterdam in 1733.[82] These texts deserve further study, but since the former has not been edited and the latter falls outside our narrow period, they are not further discussed in this book. A glimpse into the printed transmission of the *Imago mundi* family is provided in our discussion of Caxton.

In the second half of the thirteenth century, information derived from the *Imago mundi* was incorporated into a new and differently structured encyclopedic text in Brunetto Latini's *Livre dou tresor* (*c.* 1260–1265), often referred to simply as *Trésor*.[83] Rarely discussed in the context of the *Imago*

'Medieval Hebrew Translations of Philosophical and Scientific Texts: *A Chronologica Table*', in *Science in Medieval Jewish Cultures*, ed. by Gad Freudenthal (Cambridge: Cambridge University Press, 2012), pp. 17–73. For more on medieval Hebrew encyclopedias generally, see Steven Harvey, ed., *The Medieval Hebrew Encyclopedias of Science and Philosophy: Proceedings of the Bar-Ilan University Conference*, Amsterdam Studies in Jewish Thought, 7 (Dordrecht: Springer Science + Business Media, 2000).

81 The information provided here and in Table 1 in the Appendix is based partly on the work of Neubauer, 'Les traductions hébraïque', and partly on George Sarton, *Introduction to the History of Science II: From Rabbi Ben Ezra to Roger Bacon* (Baltimore: Williams and Wilkins for the Carnegie Institution of Washington, 1931), pp. 591–92, 857. Neubauer notes some variation between the printed edition and the manuscripts containing what seems to be an earlier redaction, and the choice has been made here to distinguish between the two as different translations. See also references in Robert Singerman, *Jewish Translation History: A Bibliography of Bibliographies and Studies* (Amsterdam: John Benjamins, 2002), p. 96 item 679 and Gustavo Sacerdote, 'The Ninth Mehabbereth of Emanuele da Roma and the *Tresor* of Piere de Corbiac', *Jewish Quarterly Review*, 7 (1895), 711–28 (p. 711 n. 2). A lack of editions and studies dedicated specifically to these works results in uncertainty as to the date of translation, authorship, and nature of the translation, particularly for the earlier text, which may be a translation of the *Image du monde* rather than the *Imago mundi* (how close a translation is unknown, pending an edition and specialised investigation). For instance, the Hebrew translation is assumed to be based on Gossouin's text and a date of 1280 is given for it, but with no references to the source of this information in Armstrong et al., *Knowing Poetry*, p. 105. The attribution was made by Neubauer, 'Les traductions hébraïques', pp. 136–39.

82 See Abraham Melamed, 'The Hebrew Encyclopedias of the Renaissance in *The Medieval Hebrew Encyclopedias of Science and Philosophy*, ed. Harvey, pp. 441–64 (p. 444); and also references in Singerman, *Jewish Translation History*, p. 96 under item 679 and Sacerdote, 'The Ninth Mehabbereth of Emanuele da Roma', p. 711 n. 2. The book is available on Project Gutenberg at http://www.gutenberg.org/ebooks/61024 (accessed 4 July 2024).

83 To quote Alison Cornish, this is 'the most important text written by an Italian in French'; Cornish, *Vernacular Translation in Dante's Italy: Illiterate Literature* (Cambridge: Cambridge University Press, 2011), p. 75. For editions, see Brunetto Latini, *Li Livres dou Tresor*, ed. by Francis Carmody (Berkeley: University of California Press, 1948) and, more recently, Brunetto Latini, *Tresor*, ed. by Pietro G. Beltrami, Paolo Squilacioti, Plinio Torri, and Sergio Vatteroni (Turin:

mundi tradition, primarily because of the different organisation principles that govern its structure, it has nevertheless long been acknowledged that Latini's text shows marked similarities to the *Imago mundi* tradition, suggestive of borrowing, whether directly from the Latin text or via his near-contemporary Gossouin's *Image du monde*.[84] According to Garrigues, Latini used four texts by Honorius: the *Imago mundi*, the *Elucidarium*, *De solis affectibus*, and *Summa totius*, but it is nevertheless difficult to evaluate the extent to which Latini relied on the *Imago mundi* tradition because his techniques included compilation, correction, abridgment, and combination of source material from various texts and authors.[85] In other

Einaudi, 2007). An English translation is available in Brunetto Latini, *The Book of Treasure* (*Li Livres dou Tresor*), trans. by Paul Barrette and Spurgeon Baldwin, Garland Library of Medieval Literature Series B, 90 (New York and London: Garland, 1993). For a general biography of Brunetto Latini, see Bianca Ceva, *Brunetto Latini: l'uomo e l'opera* (Milan and Naples: Riccardo Ricciardi, 1965); for a brief biographical summary in English, see David Napolitano, 'Adjusting the Mirror: A Political Remake of Brunetto Latini's *Li Livres dou Tresor*', in *The Mirror in Medieval and Early Modern Culture; Specular Reflections*, ed. by Nancy M. Frelick (Turnhout: Brepols, 2016), pp. 89–111 (pp. 90–93); for a discussion in German, see Robert Luff, *Wissensvermittlung im Europäischen Mittelalter: 'Imago mundi'-Werke und ihre Prologe* (Tübingen: Niemeyer, 1999), pp. 262–313. The *Trésor* was preceded by an earlier, shorter work, the *Tesoretto*, described by Johannes Bartuschat as the first allegorical-didactic voyage in Italian literature, characterised, however, by a lack of any vision or otherworldly elements; Johannes Bartuschat, 'La forma allegorica del "Tesoretto" e il "Dittamondo" di Fazio degli Uberti', in *A scuola con Ser Brunetto: la ricezione di Brunetto Latini dal Medioevo al Rinascimento*, ed. I. Maffia Scariati (Florence: SIMSEL–Galluzzo, 2008), pp. 417–35 (p. 417). The first two parts of the *Tesoretto*, which discuss cosmology and geography, correspond to the first two books of the *Trésor*; Bartuschat, 'La forma allegorica', p. 419. Due to spatial constraints, only the *Trésor* is discussed in this book.

84 The relationship between this text and the *Imago mundi*, and more directly to Gossouin's *Image du monde*, is briefly referred to in Bernard Ribémont, *De Natura Rerum: Études sur les encyclopédies médiévales* (Orleans: Paradigme, 1995), pp. 347, 351–52. See also Marianne O'Doherty, 'Eyewitness Accounts of "the Indies" in the Later Medieval West: Reading, Reception and Re-use (*c.* 1300–1500)' (unpublished PhD thesis, University of Leeds, 2006), p. 20; O'Doherty, *The Indies and the Medieval West: Thought, Report, Imagination* (Turnhout: Brepols, 2013), pp. 40–41. Early discussions include Thor Sundby, *Della vita e delle opere di Brunetto Latini*, trans. by R. Renier (Florence: Successori Le Monnier, 1884), pp. 82–83, 98–99, who suggests that the Brunetto Latini and Gossouin/Gautier de Metz used the same sources, or that the *Image du monde* may have been used by the former. Note that Sundby did not have access to the French text, but only to Caxton's Middle English translation. For a fuller comparison betwen sections of the *Trésor* and *Image du monde*, and a suggestion that both are based on the same Latin exemplar, see Francis J. Carmody, 'Brunetto Latini's *Trésor*: Latin Sources on Natural Science', *Speculum*, 12 (1937), 359–66, pp. 360–62; Carmody notes similarities between the *Trésor* and the *Imago mundi* on pp. 360, 362, 363, 365. For further parallels between *Trésor* and *Imago mundi*, see also Carmody, 'Latin Sources of Brunetto Latini's World History', *Speculum*, 11 (1936), 359–70 (p. 362 n. 9).

85 Marie-Odile Garrigues, 'Honorius Augustodunensis et l'Italie', *Mélanges de l'École Française de Rome : Moyen Âge, Temps Modernes*, 84 (1952), 511–30 (p. 526). For the texts, see in Yves

words: rewriting. There is also evidence of further influence of the *Image du monde* on the *Trésor* through the incorporation of passages at later dates.[86] For individual sections of the geographical description, as will be shown here, Latini's text relies heavily on Honorius. The changes he introduces are dictated by his objectives and context, and we will return to them.[87]

Trésor, Book I chapters CXXI–CXXIIII contain the geographical description of the world (the verbal *mappa mundi*).[88] The rest of the book concerns, in order: the types of knowledge; Creation; nature of angels, soul, reason, faith; the Six Ages of the World (chapters XX–XXXVIII), with reference to the Sons of Noah, and framed in the *translatio imperii* sequence following from the kingdoms of Babylon, Greece, 'women' (=Amazons), and so on; it ends with the Kings of France (chapter XXXVIII), the Prophets (chapters XXXIII–LXI), Old and New Testaments, the Saints, the Church, and Holy Roman Emperors, followed by a section on the the Nature of Things (chapter LXXXVIIII), beginning with man and animals, the roundness of the world, and the four elements (chapter CIIII marking the beginning of *Imago*

Lefèvre, *'L'Elucidarium' et les lucidaires: Contribution, par l'histoire d'un texte, à l'histoire des croyances religieuses en France au Moyen Âge*, Bibliothèque des Écoles françaises d'Athènes et de Rome, 180 (Paris: E. de Brocard, 1954), edition of the text printed as Appendix I, pp. 343–497 of the volume; *De solis affectibus* in J.-P. Migne, ed., PL 172 (Paris: Garnier, 1895), cols. 102–20; *Summa totius*, ibid., cols. 187–96. For discussion of Brunetto Latini's technique, see Bernard Ribémont, 'Brunetto Latini, le *Livre dou Tresor* et l'histoire sainte: une réception du *De ortu et obitu patrum* d'Isidore de Séville', *Cahiers de Recherches Médiévales*, 16 (2008), 135–58, and Jean Maurice, 'Brunetto Latini compilateur : "Deviser la nature des animaus" dans le *Trésor*', *Cahiers de recherches médiévales et humanistes*, 23 (2012), 173–89, esp. pp. 181–86. As observed by David J. Wallace, Latini relied on Honorius for history, Gossouin for physics, but Solinus for Geography; David J. Wallace, 'Brunetto Latini', in *Medieval France: An Encyclopedia*, ed. by William W. Kibler, Grover A. Zinn, Lawrence Earp, and John Bell Henneman, Jr., Routledge Revivals (Abingdon: Routledge, 2016), pp. 151–52 (p. 152).

86 Pietro G. Beltrami suggests that a passage on ancient philosophers between chapters 5 and 6 was a later insertion into the text from a marginal gloss (originating from Gossouin's *Image du monde*, III.12–13); 'Appunti su vicende del *Tresor*: composizione, letture, riscritture', in *L'enciclopedismo*, ed. by Picone, pp. 311–28 (p. 323). Indeed, six manuscripts of the *Trésor* contain glosses and additional material taken from the *Image du monde*; see Rita Librandi, 'Ristoro, Brunetto, Bencivenni e la *Metaura:* intrecci di glosse e rinvii tra le opere di uno scaffale scientifico', in *Lo scafale della biblioteca scientifica in volgate* (*secoli XIII–XVI*), ed. by R. Librandi and R. Piro (Florence: Sismel–Galluzzo, 2006), pp. 101–22 (p. 105). For Gossouin's text, see *L'Image du monde*, ed. by Connochie-Bourgne.

87 Another encyclopedic work in the vernacular also incorporated a section translated from the *Imago mundi* by Vivaldo Belcalzèr, who translated Book I into Mantovan dialect under the title *Mapa del mond* and incorporated it into his translation of *De proprietatibus rerum* of Bartholomaeus Anglicus. This text is not discussed in this book for reasons of space.

88 See Table of Contents in Brunetto Latini, *Li Livres dou Tresor*, ed. by Francis Carmody, p. 4.

mundi–derived content).[89] The four elements are treated in the order water–air–fire–earth in the '*mapamonde*' section. After the geographical section, the text addresses habitation, wells and fountains, fish, serpents, birds, and animals (chapters CXXV–CC). Book II is in the the mirror of princes genre, and also deals with vices and virtues, power, fortune, glory and sin.[90] The final book of the *Trésor* deals with rhetoric.

The *Trésor* enjoyed immense popularity and represents another branch on the *Imago mundi* tree, as a number of further translations (including into Italian, Castilian, and Catalan) were produced.[91] This mirrors the pattern seen for Gossouin de Metz's *Image du monde* above.

Adaptations and Translated Extracts Drawn into Other Genres

In addition to translations and rewritings of the whole Book I, individual sections and extracts were also translated and incorporated, *en bloc*, into works of different genres. The floruit for this activity also appears to fall into the thirteenth century.

A Catalan translation of a section of *Imago mundi* I is preserved in the monumental world map known as the Catalan Atlas or the Ecumene Chart, by Elisha ben Abraham Cresques *c*.1375 and gifted by the house of Aragon to the King of France.[92]

In the mid-thirteenth century, the Flemish author Jacob van Maerlant incorporated a lengthy geographical description translated from the *Imago mundi* into his description of Darius's tomb in *Alexanders geesten* (1257–1260), his Middle Dutch translation of Walter of Châtillon's *Alexandreis* (1170–1180).[93]

89 For more on the *translatio imperii* see p. 56, 137–38; for more on the Sons of Noah, pp. 46–47 n. 33.

90 For discussion of the *Trésor* as a mirror of princes, see Napolitano, 'Adjusting the Mirror'.

91 Dawn Bratsch-Prince, 'Textual History of *Li livres dou tresor*: Fitting the Pieces Together', *Manuscripta* 31 (1993), 276–89; Silvia Tita, 'Brunetto Latini's *Li livres dou Tresor*: The Translation of a Political System', *Linguistic and Philosophical Investigations*, 13 (2014), 288–96; for the Catalan text see Curt J. Wittlin, *Llibre del tresor: versió catalana de Guillem de Copons*, 4 vols (Barcelona: Barcino, 1980–89). For the Castilian text, see Biglieri, *Las ideas geográficas*, pp. 48–50.

92 Katrin Kogman-Appel, *Catalan Maps and Jewish Books: The Intellectual Profile of Elisha ben Abraham Cresques (1325–1387)* (Turnhout: Brepols, 2020), pp. 21–22; see also references in p. 58 n. 24 above.

93 Jacob van Maerlant, *Alexanders geesten*, ed. by Johannes Franck, (Groningen: J. B. Wolters, 1882) book VII, ll. 797–1828; Berendrecht, *Proeven van bekwaamheid*, pp. 37–47; van Oostrom, *Maerlants wereld*, pp. 152–62. See also Maerlant, *Alexanders geesten*, ed. by Franck, pp. xxxvi–xliv; Margriet Hoogvliet, *Pictura et scriptura: textes, images et herméneutique des mappae mundi*

Another marginal member added to this family in the fourteenth century is Fazio degli Uberti's *Dittamondo* (*c.* 1345–1367), a very popular narrative in its time, although it was never completed.[94] It is usually considered to be a vision narrative modelled on Dante's *Commedia*, though Johannes Bartuschat argued that it took the construction of Brunetto Latini's *Tesoretto* (composed in Italian and not to be confused with the *Trésor*) as its model.[95] This puts it alongside other texts of the *Imago mundi* tradition as a didactical work.[96] There is some debate as to whether, and how much, this text relied on Honorius.[97]

In the mid-fifteenth century a Castilian translation of parts of *Imago mundi* was incorporated in the *Laberinto de Fortuna* of Juan de Mena (1411–1456).[98] Thus, in Spain, as in Italy, France, and Wales, *Imago mundi* had been used and re-used by multiple authors independently from each other, creating a complicated network of interrelated texts.[99]

(*XIIIe–XVIe siècle*) (Turnhout: Brepols, 2007), p. 94. For a brief introduction to Maerlant and his opus, see A. A. M. Besamusca, 'Courtly Literature in the Low Countries and Germany: Jacob van Maerlant and Rudolf von Ems', in *Dutch and Flemish Literature as World Literature*, ed. by T. D'haen (New York: Bloomsbury Academic, 2019), pp. 44–53; for more on Maerlant and his text, see Frits van Oostrom, *Maerlants wereld* (Amsterdam: Prometheus, 1996); van Oostrom and C. H. J. M. Kneepkens, 'Maerlants *Alexanders geesten* en de *Alexandreïs:* een terreinverkenning', *De niewe taalgids*, 69.6 (1976), 483–500; for a discussion of his use of *Imago mundi*, see Petra Berendrecht, *Proeven van bekwaamheid: Jacob van Maerlant en de omgang met zijn Latijnse bronnen* (Amsterdam: Prometheus, 1996), pp. 37–68.

94 Achille Pellizzari, *Il Dittamondo e la Divina Commedia: Saggio sulle fonti del 'Dittamondo' e sulla imitazione dantesca nel secolo xiv* (Pisa: Tipografia Editrice Francesco Mariotti, 1905), p. 10; and Dario del Puppo, 'Fazio degli Uberti', in *Medieval Italy: An Encyclopedia*, ed. by Christopher Kleinhenz (New York and London: Routledge, 2004), pp. 334–35 (p. 334).

95 Surviving in over 60 manuscripts, it was the subject of an early-fifteenth-century commentary; Bartuschat, 'La forma allegorica', pp. 425–26.

96 Bartuschat, 'La forma allegorica', p. 428.

97 Early studies addressing the subject of the sources of the *Dittamondo* are Pellizzari, *Il Dittamondo* and Luigi Zambarelli, *L'imitazione dantesca nel 'Dittamondo' e nelle liriche morali di Fazio degli Uberti* (Rome: Atena, 1942). Identification of degli Uberti's sources is complicated by his tendency to condense information; Nadia Belliato, 'Per un commento al "Dittamondo": il Paradiso terrestre e la personificazione di Roma (i.xi)', *Quaderno di italianistica 2015*, Sezione di Italiano dell'Università di Losana (Pisa: ETS, 2015), 33–48 (p. 34). These issues are discussed further in Chapter 5.

98 Mena is often seen as standing between the Middle Ages and the Renaissance, but belonging more to the former; see e.g. Frank Domínguez, 'Juan de Mena's *El Laberinto de Fotruna*, Petrarch's Africa, the Scipio/Caesar Controversy', *La Crónica*, 42 (2014), 137–68 (p. 137) and Chandler Rathfon Post, 'The Sources of Juan de Mena', *Romanic Review*, 3 (1912), 223–79 (p. 251). For a discussion of the geographical passages in the *Laberinto* as a partial translation of the *Imago mundi*, see Petrovskaia, 'The Limits of Medieval Translation: Juan de Mena's Use of the *Imago mundi* in the *Laberinto de Fortuna*'.

99 A glossator of Juan de Mena's text in Palma de Mallorca, Biblioteca de Fondación Bartolomé March, MS B80-B-17 (*c.* 1470–1480) later quoted the *Imago mundi* extensively alongside Isidore's

Mention must also be made here of the Old Norse extracts surviving in the form of geographical descriptions in the sagas. This evidence supplements that of the translated fragments already mentioned. A very brief passage of European focus, which has been identified as derived from the *Imago mundi*, features in the Old Norse context in the *Fóstbræðra saga*, dated to the thirteenth century.[100]

The geographical structure of the world represented in *Eiriks saga víðförla* corresponds to that of the *Imago mundi* tradition, as the Earthly Paradise is located in the Far East, beyond India; according to Jensen, the first part of the saga contains paraphrases from both the *Imago mundi* and the *Elucidarius*.[101]

Etymologies and *Sententiarum sive de summo bono libri III* in his glosses to the *Laberinto de Fortuna*; P. A. M. Kerkhof, 'Las glosas al *Laberinto de Fortuna* en los mss. PN7 y PMM1 (*olim* MM1)', in *Juan de Mena: De letrado a poeta*, ed. by Cristina Moya García, (Cambridge: Boydell & Brewer, 2015), pp. 217–54 (p. 241). It is possible that other *Imago mundi*–derived glosses survive in different contexts that still await identification and cataloging.

100 This is a description of Rome, giving its origins, association with the lion, and etymology for the Lateran; David McDougall, 'A Note on the Shape of Rome in the *Fóstbræðra saga*', *Gripla*, 6 (1984), 259–64 (quoting the passage on pp. 259–60). This clearly echoes the Latin, but adds other information (a reference to Romulus), implying alternative or intermediate sources. For the saga, see Jacoba M. C. Kroesen, *Over de compositie der Fóstbræðra saga* (Leiden: Universitaire Pers, 1962); Jónas Kristjánsson, 'Elements of Learning and Chivalry in *Fóstbræðra saga*', in *The Icelandic Sagas and Western Literary Tradition: Proceedings of the First International Saga Conference, University of Edinburgh, 21–29 August, 1971*, ed. by Hermann Pálsson and Desmond Slay (London: Viking Society for Northern Research, 1973), pp. 259–99; Preben Meulengracht-Sørensen, 'Modernitet og tratitionalisem: et bidrag til islændingesagaernes litteraturhistorie med en diskussion af *Fóstbræðra sagas* alder', in *Die Aktualität der Saga: Festschrift für Hans Schottmann*, ed. by Stig Toftgaard Andersen (Berlin: De Gruyter, 1999), pp. 149–62. The *Imago mundi* extract is found at ch. 32 in the saga in Reykjavik, Árni Magnússon Institute, AM 142 fol (1690–1697) and AM 566 c 4to (Iceland, 1705). The passage is identified and discussed in McDougall, 'A Note'. For the manuscripts, see the catalogue entries at https://handrit.is/en/manuscript/view/is/AM02-0142 (accessed 3 April 2023), and https://handrit.is/en/manuscript/view/is/AM04-0566c (accessed 3 April 2023); also Kroesen, *Over de compositie der Fóstbræðra saga*, pp. 6–7.

101 Jensen discusses quotations from *Imago mundi* and *Elucidarium* in *Eiríks saga víðförla*, ed. by Helle Jensen (Copenhagen: Reizel, 1983), pp. xxx–xxxiii (*Elucidarium*), and pp. xxxvi–xli (*Imago mundi*); see also further discussion in Rudolf Simek, 'Die Quellen der *Eiríks saga víðförla*', *Skandinavistik: Zeitschrift für Sprache, Literatur und Kultur der Nordischen Länder*, 14 (1984), 109–14 (pp. 109–10) (*Elucidarium*) and pp. 110–11 (*Imago mundi*); and Jensen, '*Eiríks saga*', p. 161; Margaret Schlauch, *Romance in Iceland*, 2nd edn (New York: Russell & Russell, 1973), p. 49. The image of India in this saga is examined in Claire Musikas, 'A Road to India: The Example of *Eireks saga víðförla*', *Brathair*, 11 (2011), 95–104. See also Sverrir Jakobsson, 'On the Road to Paradise: "Austrvegr" in the Icelanding Imagination', in *The Fantastic in Old Norse/Icelandic Literature: Sagas and the British Isles. Preprint papers of the 13th international Saga Conference, Durham and York, 6th–12th August, 2006*, ed by John McKinnell, David Ashurst, and Donata Kick (Durham, University of Durham, 2006), pp. 935–43, and David Ashhurst, 'Imagining Paradise', in *The Fantastic in Old Norse/Icelandic Literature*, ed. by McKinnell et al., pp. 71–80 (pp. 78–79).

As none of the direct quotations identified by Jensen in *Eiriks saga* have a direct bearing on Europe, we shall not dwell on these in detail. However, it is important to note that the apparent incorporation of material from the *Imago mundi* and possibly its vernacular translations, alongside other sources into the production of local narrative literature fits the pattern seen in other branches of the tradition (French, Anglo-Norman, Welsh, Spanish), where multiple parallel rewritings (adaptations and translations) of the Latin text into the vernacular coexist and attest to the wide circulation of the Latin text, in addition to the evidence afforded by surviving manuscripts of *Imago mundi* from those areas.

Adaptation into Latin

The picture is complicated even further by the fact that in line with the medieval tendency to translate within as well as across languages, 'translation' of *Imago mundi* material did not always cross linguistic boundaries, and it was also used in a number of later Latin texts.[102] For instance, the *Imago mundi* was used extensively by Gervase of Tilbury (*c.* 1165–*c.* 1234) in his *Otia imperialia* (*c.* 1209–1215), composed for the future Otto IV.[103] The degree to which geography is incorporated in that work, known originally as *Liber de mirabilibus Mundi*, has led to extensive arguments about the relationship between Gervase and the Ebstorf Map.[104] Though the arguments, which relate to the date of the map primarily, lie beyond the range of the present study, it is worth noting that the debate illustrates yet again the complicated interrelationships between the visual *mappae mundi* tradition

102 Ellis, 'Translation and Frontiers in Late Medieval England: Caxton, Kempe, and Mandeville', p. 562.

103 Flint, ed., *Imago mundi*, pp. 11–12; O. Doberentz, 'Die Erd- und Völkerkunde in de Weltchronik des Rudolf von Hohen-Ems', *Zeitschrift für deutsche Philologie*, 12 (1881), 387–454 (pp. 412–18); Fortunata Latella, 'Come lavorava un intelettuale laico del medioevo: Gli *Otia imperialia* di Gervasio di Tilbury tra *intenvio* e *compilatio*', *Revista de literatura medieval*, 25 (2013), 103–34 (pp. 109, 113, 116); Jerzy Strzelcsyk, 'Gervase of Tilbury (*c.* 1165–*c.* 1234)', in *Trade, Travel, and Exploration in the Middle Ages. An Encyclopedia*, ed. by John Block Friedman and Kristen Mossler Figg, with Scott D. Westrem and Gregory G. Guzman (New York and London: Routledge, 2000), pp. 227–28; from Leszek Wojciechowski, 'Geographical Descriptions of Spain in Latin Medieval Encyclopedias: From Isidore of Seville to Vincent of Beauvais (VII-XIII C.)', *Roczniki Humanistyczne*, 67 (2019), 31–55 (p. 49). For the text, see Gervase of Tilbury, *Otia Imperialia: Recreation for an Emperor*, ed. and trans. by S. E. Banks and J. W. Binns (Oxford: Clarendon Press, 2002); see p. xxxix for the date of the text.

104 Armin Wolf, 'The Ebstorf *Mappamundi* and Gervase of Tilbury: The Controversy Revisited', *Imago mundi*, 64 (2012), 1–27.

and the textual tradition of the *Imago mundi* texts, on the one hand, and the transmission and translation of texts within the Latin language environment across Europe, on the other.

Thomas of Cantimpré may also have used the *Imago mundi*, as did Michael Scot (Scotus) (1175–*c.* 1232).[105] Godfrey of Viterbo (*c.* 1120–*c.* 1196) also cited Honorius as an authority, referring to him by name, and appears to quote from the third book of the *Imago mundi*.[106] Vincent of Beauvais refers to the *Imago mundi* in the *Speculum naturale*, and it was one of the geographical sources used by John of Saint-Victor in the geographical compilation within the *Memoriale historiarum*.[107] Finally, though not a direct translation of Honorius's *Imago mundi*, the *Ymago mundi* of Cardinal Pierre d'Ailly/Pietrus Aliacus (*c.* 1350–*c.* 1420). exclusively geographical in nature, borrowed heavily from Book I of its twelfth-century predecessor.[108] Indeed, various passages in the *Ymago mundi* have been identified as direct borrowings or quotations from *Imago mundi*.[109] His reliance on Honorius for discussion

105 Flint, ed., *Imago mundi*, pp. 11–12; Schorbach, *Studien*, p. 162. The use of the *Imago mundi* as a source by Michael Scot for the discussion of the music of the spheres in *Liber quatuor distinctionum* is noted by Christian Meyer, 'Musique et astronomie dans le *Liber quatuor distinctionum* de Michael Scot', *Archives d'histoire doctrinale et littéraire du Moyen Âge*, 76 (2009), 119–77 (pp. 135–36). As Michael Scot's use does not include the Europe section of the *Imago mundi*, his work is not discussed further in this book.

106 Quoted and discussed in Klaus Oschema, *Bilder von Europa im Mittelalter*, Mittelalter-Forschungen, 43 (Ostfildern: Thornbecke, 2013), pp. 338–39. Gottfried also used the *Elucidarium*; Dagmar Gottschall, 'Marius Salernitanus und Gottfried von Viterbo', *Sudhoff's Archiv*, 75 (1991), 111–13 (p. 112).

107 M. Franklin-Brown, *Reading the World: Encyclopedic Writing in the Scholastic Age* (Chicago: Chicago University Press, 2012), pp. 101–04; Vincent calls the text *Speculum mundi* (p. 273). For more on John's use of Honorius, see Nathalie Bouloux, 'From Gaul to the Kingdom of France: Representations of French Space in the Geographical Texts of the Middle Ages (Twelfth–Fifteenth Centuries)', in *Space in the Medieval West: Places, Territories and Imagines Communities*, ed. by Meredith Cohen and Fanny Madeline (Farnham and Burlington: Ashgate, 2014), pp. 197–217 (pp. 200–01).

108 More so, as Valerie I. J. Flint points out, than is noted by Buron in his edition: see Flint, *The Imaginative Landscape of Christopher Columbus* (Princeton, NJ: Princeton University Press, 1992), p. 48 n. 14, p. 54. The title is sometimes spellt *Imago mundi*; I use the spelling of the title of the edition to distinguish it from Honorius's text; for the edition, see Edmond Buron, ed. and trans., *Ymago mundi de Pierre d'Ailly*, 3 vols (Paris: Maisonneuve Frères, 1930).

109 For instance, passages regarding multicoloured islanders and giants, referred to in Chet Van Duzer, 'Cartographic Invention: The Southern Continent on Vatican MS Urb. Lat. 274, Folios 73v-74r (*c.* 1530)', *Imago mundi*, 59 (2007), 193–222 (pp. 200 and 221 n. 93). J. D. North notes that Pierre d'Ailly was 'a polemicist rather than an astronomer' and that he 'borrowed heavily from his predecessors', to such a degree as to reproduce entire sections of their writings verbatim; North, 'The Western Calendar – "Intolerabilis, Horribilis, et Derisibilis": Four Centuries of Discontent', in *Gregorian Reform of the Calendar: Proceedings of the Vatican Conference to Commemorate its*

of Europe requires a separate study, and is for now left out of the present discussion, which focuses primarily on the vernacular tradition.

The overview above has provided a brief introduction to a part of this complicated tradition. The texts of the four categories introduced in this chapter will be discussed in more detail in the following chapters, specifically in terms of their treatment of Europe. There are other works which rely on or quote *Imago mundi*, and some of these will also be mentioned. These include texts whose degree of reliance on the *Imago mundi* is difficult to establish with certainty, such as Fazio degli Uberti's *Dittamondo*. Indeed, it is possible that not all heirs of the *Imago mundi* have been identified hitherto, and it may be that future research will bring to light more connections and new patterns of influence. Even when a medieval text is used extensively as a source, not only might it not be cited, but it might not be even quoted verbatim, or it may be used only to form the template for the narrative structure.[110] Influences, therefore, are difficult to trace. This aspect of medieval translation and transmission of knowledge complicates considerably the study of the influence of texts such as the *Imago mundi*.

400th Anniversary 1582–1982, ed. by G. V. Coyne, M. A. Hoskin, and O. Pedersen (Vatican City: Specola Vaticana, 1983) pp. 75–113 (p. 88).

110 Maria Teresa Rachetta, 'Brunetto Latini, La Storia Universale e la letteratura francese di matrice erudita del primo xiii secolo', in *Dante e la cultura fiorentina. Bono Giamboni, Brunetto Latini e la formazione intellettuale deil laici*, ed. by Zygmunt G. Barański, Theodore J. Chachey, Jr., and Luca Lombardo (Rome: Salerno, 2019), pp. 101–132 (p. 123).

Part II

Modes of Reading Geography

3. *Time.* Authority and Archaism

Abstract: This chapter explores the archaisms in the territorial nomenclature of the *Imago mundi.* To illustrate how this flexible framework permeated vernacular tradition, the chapter begins with an examination of the Old Norse material. The Norse evidence for transmission of the *Imago mundi* and the incorporation of its material into new texts of varied genres, exemplifies the use and re-use of the same material in various contexts typical of the entire tradition. The second and third sections of the chapter focus on Honorius's use of outdated Roman imperial nomenclature, and how later adaptors, such as Brunetto Latini and Gervase of Tilbury exploit it by overlapping new political maps onto this base description, using it, in essence, as a base outline.

Keywords Scandinavia; Roman Empire; geographical archaism; imperial territorial units; *provincia*; ovelapping maps

Chapters 1 and 2 have provided an overview of the content and the spread of the *Imago mundi.* Now that we have gathered our affairs, we can mount the hippogriff and in Chapter 3 set out on our journey, as a number of important ideas that have surfaced in the initial chapters are brought together. We will examine how the *Imago mundi* becomes an authority drawn on for geographical information in different cultural and linguistic concepts. This discussion builds on the treatment in Chapter 2 which touched briefly on medieval ideas of translation and authority.[1] We then turn to the first of two major aspects of the text that I argue are crucial to its transformation into an authority and fenomenal success: archaism. (The second is the flexibility of the spatial description, which I discuss in terms of fuzzy sets in Chapter 4). The topographical framework in the *Imago mundi,* I will argue, like the reference to the Pillars of Hercules in *Orlando Furioso,* provides classical landmarks as reference points for medieval geography.

1 Chapter 2, pp. 55–57.

Petrovskaia, N.I. *Transforming Europe in the Images of the World, 1110–1500. Fuzzy Geographies.* Taylor & Francis Group, 2025
DOI 10.5117/9789048563166_CH03

The nomenclature used by Honorius is predominantly that of Roman provinces. This is sometimes updated in vernacular rewritings. The phenomenon particularly concerns European regions, and particularly in relation to areas close to the place of production for the updated version or locations to which the author or audience of the new vernacular text had particular connections.[2] Reference is also sometimes made to contemporaneous political entities, and these cases will recieve our special attention. I argue here that this is evidence for the remarkable strength of the notions of authority in the medieval period, coupled with a reluctance to challenge that authority through change motivated by anything less than equal authority (general knowledge, it seems, would have been accepted).[3] One might run into trouble only, as Marco Polo almost did, when the evidence of a unique eye-witness (himself, in Marco Polo's case) ran contrary to widely circulated authorities.[4] For the depiction of Europe in the texts of the *Imago*

2 This theme is built upon in Chapter 5, where we look at the context of our texts: audience experience and patronage.

3 For the acceptability of expectations of general knowledge as reason for adjustment of information, one need only think of Caxton's appeals to his audience in the few instances where he made explicit alterations to the text of the *Image du monde* which he was translating; see James A. Knapp, 'Translating for Print: Continuity and Change in Caxton's *Mirrour of the World*', *Disputatio*, 3 (1998), 64–90 (pp. 73–75). As Knapp observes, 'Caxton is careful to reserve his alterations for aspects of the French text that are obviously wrong (confirmed so by personal experience) or obviously offensive to an English audience' (p. 74). Knapp's comment about personal experience relates to Caxton's removal of the claim made in the French text that Englishmen have tails. For the sake of the present argument it seems appropriate to point out that the experience would have been shared by Caxton's audience. Caxton changed the text because his audience would have known as well as he did that it was wrong (and, of course, as Knapp points out, would have found this offensive). The fact that Caxton often retains the error alongside his corrections (which then function as a guide) suggests that not only does the translator/editor in this case function as a guide to the readers, as Knapp observes, but also that there is complicity between the editor who provides the correction and the reader, whom the editor expects to agree with this correction.

4 For the negotiation between authority of tradition and authority of the eyewitness in Marco Polo's text, see, for instance, discussions in Aníbal A. Biglieri, *Las ideas geográficas y la imagen del mundo en la literatura española medieval*, Medievalia Hispanica, 17 (Madrid: Iberoamericana; Frankfurt am Main: Vervuert, 2012), pp. 25–30; Simon Gaunt, *Marco Polo's 'Le Devisement du Monde': Narrative Voice, Language and Diversity* (Cambridge: D. S. Brewer, 2013), pp. 30–35, 113–44; Campbell, *The Witness and the Other World: Exotic European Travel Writing, 400–1600* (Ithaca, NY and London: Cornell University Press, 1988), pp. 94–112; Suzanne M. Yeager, 'The World Translated: Marco Polo's *Le Devisement dou monde*, *The Book of Sir John Mandeville*, and their Medieval Audiences', in *Marco Polo and the Encounter of East and West*, ed. by Suzanne Conklin Akbari and Amilcare Iannucci, with the assistance of John Tulk (Toronto, Buffalo, and London: University of Toronto Press, 2008), pp. 156–81 (pp. 156, 167–70). For a more general discussion of the relation between eyewitness accounts and textual authority, see Bernard Ribémont,

mundi tradition, in the context of a culture used to perceiving the world as a synchronous representation of time-space (which Georg Jostkleigrewe called the historico-geographical model), so vividly illustrated by the *mappae mundi*, this did not represent a conflict.[5] This framework provides ample material for rewriting and renegotiation, especially, as we shall see in later chapters, of one's own place in the world.

To illustrate how intensively this flexible framework permeated vernacular tradition, it is useful to look at the Norse material, briefly introduced in the previous chapter, in a little more detail. Even though – perhaps even because – it is fragmentary and circumstantial, the Norse evidence for transmission of the *Imago mundi* and the incorporation of the material derived from it into new texts of widely varied genres, exemplifies the use and re-use of the same material in various contexts typical of this whole tradition.[6]

Authority in Fragments

Scandinavia barely features in the *Imago mundi*: 'In hac est Dania et Norweia' ('In that [= *Germania Inferior*] is Denmark and Norway'; 1.24) is all the text says of it.[7] Previous discussions have shown that echoes of this geographical tradition, and the tripartite world structure in particular can be found throughout Norse literature.[8] *Imago mundi* is cited and quoted in multiple

'L'inconnu géographique des encyclopédies médiévales', *Cahiers de recherches médiévales*, 3 (1997), 101–11, esp. pp. 101–03, and Surekha Davies, 'The Wondrous East in the Renaissance Geographical Imagination: Marco Polo, Fra Mauro and Giovanni Battista Ramusio', *History and Anthropology*, 23 (2012), 215–34 (pp. 220–23 and 226–29).

5 See, for instance, Naomi Reed Kline's discussion of the Hereford *mappa mundi* as a 'universal chronicle'; Kline, *Maps of Medieval Thought: The Hereford Paradigm* (Woodbridge: Boydell, 2001), pp. 221–22; also Evelyn Edson, *The World Map, 1300–1492: The Persistence of Tradition and Transformation* (Baltimore: Johns Hopkins University Press, 2007), pp. 22–24. Jostkleigrewe, 'L'espace entre tradition et innovation: la géographie symbolique du monde et son adaptation par Gossouin de Metz', *Construction de l'espace au Moyen Age: Pratiques et représentations. XXXVIIe Congrès de la SHMES. Mulhouse, 2–4 juin 2006* (Paris: Publications de la Sorbonne, 2007), pp. 369–78 (p. 377). See discussion in Chapter 3, p. 111.

6 Margaret Clunies Ross and Rudolf Simek, 'Encyclopedic Literature', in *Medieval Scandinavia: An Encyclopedia*, ed. by P. Pulsiano, Kirsten Wolf, Paul Acker, and Donald K. Fry (New York and London: Garland, 1993), pp. 164–66 (p. 166).

7 Honorius Augustodunensis, *Imago mundi*, ed. by Valerie I. J. Flint, *Archives d'histoire doctrinale et littéraire du Moyen Âge*, 49 (1982), p. 60. My translation.

8 A number of relevant studies are referenced throughout the following discussion. For the most recent overview and further bibliography, see Dale Kedwards, 'Geography', in *A Critical*

surviving medieval sources.[9] Indeed, Paul Lehmann has compared the influence of the *Imago mundi* to that of Honorius's *Elucidarium* (a full translation of which into Icelandic does survive), and even that most influential of medieval encyclopedias, Isidore's *Etymologies*.[10] It has been argued that influence of the *Imago mundi* on Old Norse literature was formative for the development of Norse understanding of space in general.[11] This provides context for the surviving echoes of the geographical content of the *Imago mundi* in the Norse sagas.[12] The texts that cite it include *Stjórn* (a Norwegian biblical compendium); *Eiriks saga víðförla* ('Saga of Erik the Far-Travelled'), which concerns the quest for the Earthly Paradise; *Fóstbrœðra saga* ('Saga of the Sworn Brothers'); and the *Dinus saga drambláta* ('Saga of Dinus the Proud').[13] It has also been argued that the *Imago mundi* had influenced the

Companion to Old Norse Literary Genre, ed. by Massimiliano Bampi, Carolyne Larrington, and Sif Rikharðsdóttir (Cambridge: D. S. Brewer, 2020), pp. 127–44.

9 Rudolf Simek, *Altnordische Kosmographie: Studien und Quellen zu Weltbild und Weltbeschreibung in Norwegen und Island vom 12. bis zum 14. Jahrhundert* (Berlin and Boston: De Gruyter, 2012), p. 27.

10 Lehman, 'Skandinaviens Anteil an der Lateinischen Literatur und Wissenschaft des Mittelalters', in *Erforschung des Mittelalters: Ausgewählte Abhandlungen und Ausätze*, 5 vols (Stuttgart: Anton Hiersemann, 1959–73), vol. 5 (1962), pp. 275–393 (p. 358). For more on the Old Norse *Elucidarium*, see above, p. 43, n. 21 and discussion also in G. Turville-Petre, *Origins of Icelandic Literature* (Oxford: Clarendon Press, 1953), pp. 137–40.

11 Lukas Rösli points out that the work of Rudolf Simek and Margaret Clunies Ross has demonstrated that 'Old Norse cosmographical and spatial narratives, on the one hand, did not occur autonomously, but were dependent on contemporary scholarly discourses from Europe, and, on the other hand, that these texts shaped at least as much the cultural and collective understanding of space in medieval Iceland as they were in turn influenced by a conception of space'; Rösli, 'Spatial Studies', in *Handbook of Pre-Modern Nordic Memory Studies: Interdisciplinary Approaches*, ed. by Jürg Glauser, Pernille Hermann, and Stephen A. Mitchell, 2 vols (Berlin: De Gruyter, 2018), I, pp. 274–83 (p. 277). A similar pattern was also observed for cartography by Dale Kedwards in *The Mappae Mundi of Medieval Iceland* (Woodbridge: Boydell & Brewer, 2020).

12 Further context is provided by the presence translation of other works by Honorius Augustodunensis into Old Norse, including the *Elucidarium* and a sermon surviving in Copenhagen, University of Copenhagen, AM 655XX 4to (s. XIV); *'Elucidarius' in Old Norse Translation*, ed. by Evelyn Scherabon Firchow and Kaaren Grimstad (Reykjavik: Stofnun Árna Magnússonar, 1989); Stephen Pelle, 'An Old Norse Adaptation of a Christmas Sermon by Honorius Augustodunensis', *Medieval Sermon Studies*, 61 (2017), 44–58.

13 Simek, *Altnordische Kosmographie*, p. 27. See also Svanhildur Óskarsdóttir, 'Universal History in Fourteenth-Century Iceland: Studies in AM 764 4to' (unpublished PhD thesis, University of London, 2000), pp. 80–83; Ian J. Kirby, *Bible Translation in Old Norse* (Geneva: Droz, 1986), pp. 61, 64, 69–71; Frederic Amory, 'Things Greek and the Riddarasögur', *Speculum*, 59 (1984), 509–23 (p. 522). Helle Jensen, ed., *Eiríks saga víðförla* (Copenhagen: Reizel, 1983). For a brief overview, see Helle Jensen, '*Eiríks saga víðförla*', in *Medieval Scandinavia: An Encyclopedia*, ed. by Pulsiano et al., pp. 160–61.

structure of *Snorra Edda* itself, suggesting a fundamental impact on the Norse worldview.[14]

Although the use of *Imago mundi* material is extremely limited in all cases cited here, the sheer range of texts that use it attests to its status as authority.[15] This incorporation of Honorius's geographical view into the Old Norse world view (where it also becomes tied to self-representation) reflects processes that have been observed for other aspects of local literary production – for example, genealogies, which came to combine the local with the pan-European traditions, such as tracing origins back to historical figures and Trojans.[16] This snapshot of the Old Norse tradition thus illustrates some of the main points that will be made about the *Imago mundi* as authority. I argue that its enormous popularity has a lot to do with its traditional nomenclature and with the freedom it creates for subsequent adaptors, much like the pattern of the Trojan origin legends. Part of that freedom lies in the adaptability of its map, which much like the Trojan origin legends is found in its reliance on frameworks provided by classical antiquity. In the case of Honorius's text this is expressed in its resolute and intentional archaism.

The Timelessness of Archaism

In its use of classical (and thus, by the Middle Ages, outdated) place-names, the *Imago mundi* is purposefully archaising, a characteristic retained

14 Gunnar Harðarson, 'Old Norse Intellectual Culture: Appropriation and Innovation', in *Intellectual Culture in Medieval Scandinavia,* c. *1100–1350*, ed. by Stefka Georgieva Eriksen (Turnhout: Brepols, 2016), pp. 35–73 (p. 59); and Margaret Clunies Ross, 'The Influence of the Medieval Encyclopedia on Snorri's *Edda*', in *Skaldskaparmal: Snorri Sturluson's Ars Poetica and Medieval Theories of Language*, Viking Collection, 4 (Odense: Odense University Press, 1987), 151–73 (155–67); originally published in *The Sixth International Saga Conference 28/7–2/8 1985* (Copenhagen, 1985), pp. 177–206. For more on descriptions of world geography, modelled on the tripartite structure seen in *Imago mundi*, in Old Norse literary texts, see Kedwards, 'Geography', pp. 127–44.

15 In the case of *Dínus saga*, for instance, it is confined to a geographical reference to the three parts of the world in the beginning of the text. For the text, see Finnur Jónsson, ed., 'Dínus rímur', in *Rímnasafn: Samling af de aldeste islandske rimer*, 2 vols (Copenhagen: Møller & Jørgensen, 1905–1922), II, pp. 801–24; Jónas Kristjánsson, ed., *Dínus saga drambláta*, Riddarsögur, 1 (Reykjavik: Háskóli Íslands, 1960).

16 Judy Quinn, 'From Orality to Literary in Medieval Iceland', in *Old Icelandic Literature and Society*, ed. by Margaret Clunies Ross (Cambridge: Cambridge University Press, 2000), pp. 30–60 (pp. 48–49); referring to Eyvindr Fjeld Halvorsen, 'Landfreðgatal', *Kulturhistorische leksikon for nordisk middelalder*, 10 (1965), cols. 311–13.

in several of its vernacular rewritings.[17] This is unsurprising given the conservative nature of the text, which presents an accessible version of the encyclopedic status quo and relies heavily in this on the Isidoran tradition.[18] The main sources used very visibly in the *Imago mundi* are the trio of early medieval encyclopedic greats: Isidore, Orosius, and Bede. Additional sources include Pliny, Solinus, Macrobius, Martianus Capella, Rabanus Maurus, Helpericus, Hyginus, Cassiodorus, Pseudo-Bede, and Pseudo-Alcuin, among others, all notably reflective of what has been described as the 'Carolingian worldview'.[19] We will see in the following that it is precisely the fact that it was outdated from the outset that guarantees the popularity of *Imago mundi*.

As far as Honorius's geography is concerned, this archaism is most visible in the sections on Europe, where he uses predominantly Roman nomenclature (more than for other parts of the world).[20] This not only gives his text an antiquated flare, but it also sets Honorius's Europe apart from the rest of the world, particularly in his use of the term *provincia* ('province').[21] Since his use of the term is illustrative of both his reliance

17 Flint, ed., *Imago mundi*, pp. 14–18.

18 Bernard Ribémont, *Les origines des encyclopédies médiévales: D'Isidore de Seville aux Carolingiens* (Paris: Champion, 2001), pp. 234, 316; for a discussion of Honorius in the context of twelfth-century learning, see H. Schipperges, 'Honorius und die Naturkunde des 12. Jahrhunderts', *Suddhovs Archiv: Zeitschrift für Wissenschaftsgeschichte*, 42 (1958), 71–82.

19 Michael W. Twomey, 'Honorius Augustodunensis', *Trade, Travel, and Exploration in the Middle Ages: An Encyclopedia*, ed. by John Block Friedman and Kristen Mossler Frigg, with Scott D. Westrem and Gregory G. Guzman (New York and London: Routledge, 2000), pp. 259–61 (p. 260); Marie-Odile Garrigues, 'L'oeuvre d'Honorius Augustodunensis: Inventaire Critique' (1), *Abhandlungen der Braunschweigischen Wissenschaftlichen Gesellschaft*, 38 (1986), 7–136 (pp. 30–32); Flint, ed., *Imago mundi*, p. 18. For an overview of the early medieval worldview, see, for example, N. Lozovsky, *'The Earth is Our Book': Geographical Knowledge in the Latin West, ca. 400–1000* (Ann Arbor, University of Michigan Press, 2000).

20 This is arguably because knowledge of other parts of the world would have been more limited and more slowly updated. There is room here for further exploration, specifically of representations of Asia and Africa in the *Imago mundi* tradition, but these have to be left to other studies. For recent studies of medieval representations of Africa and Asia more broadly, see, for example, Francesc Relaño, *The Shaping of Africa: Cosmographic Discourse and Cartographic Science in Late Medieval and Early Modern Europe* (Aldershot: Ashgate, 2002); Suzanne Conklin Akbari, *Idols in the East: European Representations of Islam and the Orient, 1100–1450* (Ithaca, NY and London: Cornell University Press, 2009); Marianne O'Doherty, 'Eyewitness Accounts of "the Indies" in the Later Medieval West: Reading, Reception and Re-use (*c.* 1300–1500)' (unpublished PhD thesis, University of Leeds, 2006); and O'Doherty, *The Indies and the Medieval West: Thought, Report, Imagination* (Turnhout: Brepols, 2013).

21 For the term *provincia* in the classical Roman context, see John Richardson, 'Provincial Administration', in *The Oxford Handbook of Roman Law and Society*, ed. by Paul J. du Plessis, Clifford Ando, and Kaius Tuori (Oxford: Oxford University Press, 2016), pp. 111–23, (pp. 112–13); and Alejandro Díaz Fernández, ed., *Provinces and Provincial Command in Republican Rome:*

on Roman nomenclature and of the distinction between how Europe and other regions are treated, it is explored here further.

Although Honorius bases his description of Europe largely on Isidore's *Etymologies* XIV, his use of the territorial nomenclature does not correspond to Isidore's. For instance, where Isidore uses the term *regio* for Scythia, Honorius does not (he uses no territorial term for Scythia). Meanwhile where for Alania Isidore uses *pars* and for Dacia and Gothia no term at all, Honorius designates these as *provincie*.[22] Returning to Germany, Isidore designates Moesia and Pannonia as *provincie*, and Noricum as *Noricus ager*.[23] The term *ager*, originally meaning 'field' but also used to signify 'territory, district, domain, the whole of the soil belonging to a community', is not used by Honorius at all.[24] The other source for these passages identified by Flint, Osorius's *Historiae adversum paganos*, does not offer a correspondence to Honorius's nomenclature either.[25] Honorius's nomenclature thus appears to be his own.

Starting with the East, as the text does, we can observe that there is not a single instance of the use of the term *provincia* in the *Imago mundi* until

Genesis, Development and Governance (Sevilla and Zaragoza: Editorial Universidad de Sevilla and Prensas de la Universidad de Zaragoza, 2021).

22 Isidore, *Etymologies* in *Isidori Hispalensis Episcopi Etymologiarvm sive Originvm Libri xx*, ed. by W. M. Lindsay, 2 vols (Oxford: Oxford University Press, 1911), xiv.iv.2–3; Flint, ed., *Imago mundi*, I.22 (p. 59).

23 Isidore, *Etymologies*, ed. by Lindsay, XIV.iv.5.

24 Definition quoted from Charlton T. Lewis and Charles Short, *A Latin Dictionary* (Oxford: Clarendon Press, 1879), s.v. *ager*.

25 *Historiarum adversum Paganos*, I.2; see *Pavli Orosii Historiarum Adversum Paganos libri VII*, ed. by Karl Zangemeister (Vienna: apud Geroldi Filium Bibliopolam Academiae, 1882), pp. 9–10. Orosius does not use the term *provincia* for any of the regions discussed here and cannot have been the basis for Honorius's understanding of *provincia*, as in his use, provinces and countries were composed of *gentes* (and thus were larger units), and he tended to use *gentes* where his own sources might have used *provincia*; see discussion in Hans-Werner Goetz, '*Gens*: Terminology and Perception of the "Germanic" Peoples from Late Antiquity to the Early Middle Ages', in *The Construction of Communities in the Early Middle Ages: Texts, Resources, and Artefacts*, ed. by R. Corrandini, M. Diesenberger, and Helmut Reimitz (Leiden and Boston: Brill, 2003), pp. 39–64 (pp. 45–46). Goetz also presents a useful quantitative analysis of the use of territorial terminology (*fines*, *regio*, *regnum*, and *provincia*) in the early medieval period (ibid., pp. 55–57). His study, however, relates specifically to ethnonyms and does not include a differentiating semantic analysis for the territorial terms. For a discussion of the use of territorial terms in French and German medieval sources, see Thomas Kohl, 'Territorialisierungen im 11. Jahrhundert? Der deutsche Südwesten und das westliche Frankreich im Vergleich', in *Territorium. Raum und Politik: Wahrnehmung und Praxis im Frankenreich und seinen Nachfolgereichen vom 9. bis zum 13. Jahrhundert* (Tübingen, 2011), http://nbn-resolving.de/urn:nbn:de:bsz:21-opus-59604 (accessed 17 July 2020).

the reference to *provincia Thebaida* in I.17, on Egypt.[26] Until that point Honorius only uses the term *regio* ('region'), including for Iudea, even though there is a reference to the Roman Empire in that chapter, with the mention of Emperor Hadrian.[27] The use of the term for Thebaid is one of only two instances of the term 'province' in Honorius's discussion of Asia. The other is Bithynia: *prima provincia Asie Minoris est Bithinia* ('the first province of Asia Minor is Bythinia').[28] By contrast, Honorius's Europe is full of 'provinces'. He uses the term for: Alania, Dacia and Gothia (I.22); Dalmacia and Aonia (I.25); Tuscia, but presumably it applies to other Italian regions as well (I.26); and six provinces in Spain (I.28).[29] In his description of Africa, Lybia (I.30), Mauretania Stifensis, Mauretania Caesarensis, and Mauretania Tinguitana (I.31) are also termed provinces.[30] Although care should be exercised in drawing conclusions on the basis of the term's absence from descriptions of certain parts of Europe, some tentative observations can be made on the basis of this evidence.

To start with the least controversial of these, we can say that the use of the term *provincia* in the *Imago mundi* roughly corresponds to the extent of the Roman Empire. This is unsurprising, given Honorius's use of the names of Roman provinces even in cases where this represents serious archaism in the twelfth-century context of this work. The term is not used at all in descriptions of Germany and France (I.23–24 and I.27), although Roman toponyms are used throughout these chapters. While it is not the objective here to argue for a sharp and precise differentiation between the terms *provincia* and *regio* throughout medieval discourse, given the pattern described above, the possibility that Honorius at least made such a distinction in this particular text is worth investigating further.[31] Philippe Leveau's justification of a similar investigation undertaken in relation to classical texts also applies here: 'un tel inventaire est susceptible d'éclairer le

26 Flint, ed., *Imago mundi*, p. 57.

27 Chapter 15: Flint, ed., *Imago mundi*, p. 56.

28 Flint, ed., *Imago mundi*, p. 58; my translation.

29 Flint, ed., *Imago mundi*, pp. 59–62.

30 Flint, ed., *Imago mundi*, p. 63. Note that Tinguitana is also in the list of Spanish provinces. This is arguably a case of overlap between two fuzzy sets – Europe and Africa.

31 For a general discussion of this terminology, see Otto Brunner, *Land and Lordship: Structures of Governance in Medieval Austria*, trans. by Howard Kaminsky and James van Horn Melton (Philadelphia: University of Pennsylvania Press, 1992), pp. 159–60; Brunner considers that these terms are used interchangeably and 'sometimes used generally for any geograpical unit' (p. 160). For a detailed study of the terms in classical sources, see Philippe Leveau, '*Territorium Urbis*: Le territoire de la cité romqine et ses divisions: du vocabulaire aux réalités administratives', *Revue des Études Anciennes*, 95 (1993), 459–71.

débat et de limiter les discussions nées des différences de sens données aux mêmes mots' ('such an inventory may throw light on the debate and limit the discussions born from different meanings given to the same words').[32]

Although we do not have an identified written source for Honorius's use of *provincia* (and as pointed out above it is in any case not Isidore), it could be a product of general knowledge, and Honorius's own contribution to the text. In each case where Honorius uses the term *provincia*, it is applied to the name of a Roman province, and not once is the term used for territories outside the Roman Empire.[33] It may be relevant to consider here the point raised by Ronnie Ellenblum in a discussion of whether borders existed in the Middle Ages, when he mentions Benjamin Isaac's argument that the only 'clearly demarcated boundaries' in the Roman Empire were those of the individual provinces.[34] One could suggest, given the evidence of the *Imago mundi*, that those provincial boundaries were the only ones that remained, at least in the learned geographical discourse (if not necessarily physically on the ground). The remaining question, then, carried over from the discussion in the previous chapter, is why the term is not used at all in the chapters describing French and German regions in Honorius's Latin text?

The proposed answer lies, despite the apparent archaism of Honorius's geographical description, in the contemporary political map of the area, for here we find one of very few instances of place-names being updated, with Noricum glossed as *Bawaria*, followed by the much-discussed reference to Honorius's own purported home of *Ratispona* (=Regensburg).[35]

Regardless of whether the *Imago mundi* was written, as Flint suggests, for Henry the Black, duke of Bavaria, Honorius's associations with Regensburg make it likely that he would have been aware of recent political developments in the region. It makes sense, therefore, that even if he designated this region by its Roman provincial name Noricum (*Noricus* in the *Imago mundi*), as

32 Leveau, '*Territorium urbis*', p. 460. My translation.

33 See above, pp. 89–90.

34 Ronnie Ellenblum, 'Were there Borders and Borderlines in the Middle Ages? The Example of the Latin Kingdom of Jerusalem', in *Medieval Frontiers: Concepts and Practices*, ed. by David Abulafia and Nora Berend (New York and Abingdon: Routledge, 2016), pp. 105–20 (p. 106); referring to Benjamin Isaac, *The Limits of Empire: The Roman Army in the East* (Oxford: Clarendon Press, 1990), p. 3.

35 For the association of Honorius with Regensburg, see Flint, ed., *Imago mundi*, p. 8; Eva Mathews Sanford, 'Honorius *Presbyter* and *Scholasticus*', *Speculum*, 23 (1948), 397–425, esp. pp. 398–403; Loris Sturlese, 'Zwischen Anselm und Johannes Scotus Eriugena: Der seltsame Fall des Honorius, des Mönchs von Regensburg', *Historia philosophiae medii aevi: Studien zur Geschichte der Philosophie des Mittelalters*, ed. by B. Mojisch and O. Pluta (Amsterdam: John Benjamins, 1991), pp. 927–50 (esp. p. 930).

well as by its current name, *Bawaria*, he might have chosen to avoid the term *provincia* on purpose.[36] The fact that in this instance Honorius provides a contemporary term alongside the Roman designation supports this theory. According to Klebel, in classical geography, the province Noricum corresponded to Carinthia (German: Kärnten) and Styria (German: Steiermark), but the term was also applied to the region on the Danube, from the river Inn to the Vienna Woods, including Salzburg and the Bavarian Alpine Foreland, though in the Middle Ages it is used primarily for the Duchy of Bavaria.[37] The exact localisation is not provided in the *Imago mundi*, which merely lists Bavaria as one of the regions in *Germania Superior*: 'Est in ea Noricus que et Bawaria, in qua est civitas Ratispona' ('There is in it Noricum, which is also Bawaria, in which is the city of Regensburg').[38] Although the sentence immediately preceding described the Danube, it is not specified how the region *Noricus*/Bavaria relates to the river. A little earlier in the same chapter, Swabia (*Suevia*) is referred to as a *regio*, which can also mean 'province' but might be regarded as a more neutral term.[39] Honorius might have avoided the term *provincia* in his description of these areas because in contemporary parlance it had acquired a new meaning, which may have been foremost in the minds of his local audience when he was writing and would have overshadowed any archaic imperial connotations.[40]

In 1098, to quote Benjamin Arnold's description of the events, 'Swabia was deprived of its ancient constitution as a *gens* under the presidency of its duke, and as a *provincia* with known geographical boundaries'.[41] It may be therefore that Honorius avoided using the term in a context where it might have been perceived to have contemporary political connotations.

36 Flint, ed., *Imago mundi*, p. 60. For the use of *Noricum* to designate Bavaria in the Middle Ages, see Ernst Klebel, 'Das Fortleben des Namens "Noricum" im Mittelalter', *Carinthia I. Mitteilungen des Geschichtsvereins für Kärnten, Klagenfurt*, 146 (1956), 481–92.

37 Klebel, 'Das Fortleben', p. 481.

38 Flint, ed., *Imago mundi*, p. 60; *ea* probably refers to *Germania Superior* mentioned at the start of the chapter (p. 59).

39 See above, p. 90.

40 Gleb Schmidt suggests the composition of the *Elucidarium* is associated with the monastic reform movement that started in Hirsau Abbey (on the Swabia/Franconia border); 'From Manual to Best-Seller: The History of Honorius Augustodunensis's *Elucidarium*', in *Books of Knowledge in Late Medieval Europe: Circulation and Reception of Popular Texts*, ed. by Pavlína Cermanová and Václav Žůrek, Utrecht Studies in Medieval Literacy, 52 (Turnhout: Brepols, 2021), pp. 137–63 (pp. 141–43). Evidence seems thus to point increasingly to Honorius's association with South-Western Germany generally.

41 Arnold, *Princes and Territories in Medieval Germany* (Cambridge: Cambridge University Press, 1991), p. 96. A *post quem* of 1098 can be given for Honorius's earliest known work, the *Elucidarium*; Schmidt, 'From Manual to Best-Seller', p. 140.

In turn, it may also explain why he may have avoided the term *provincia* for Germanic regions altogether. His only nod towards the contemporary reality would have been in providing a gloss for *Noricus* in the term *Bawaria*, which is not derived from classical Roman geography. This constitutes the only deviation from the otherwise resolutely antiquarian form of the text. It is striking, therefore, that this deviation concerns the region where the author himself was probably active. We will see throughout the discussion in the following chapters that this is a phenomenon recurring throughout the entirety of the *Imago mundi* tradition.

Both the archaism of Roman imperial terminology and the special treatment of Germany show that Honorius's map seem to derive from a socio-cultural, or even political history of the region. However, shifts in terminology observed above do not map perfectly onto what we might imagine to be boundaries that Honorius demarcates for the regions in his text (for Europe: the Don and the Mediterranean). The Roman Empire extended beyond 'Europe' and so does its provincial terminology in his use. Chapter 4 pursues this line of thought further, suggesting that in the *Imago mundi* we might not be dealing the schematic linear boundaries of regions as all, but rather with what are essentially fuzzy sets, which allow for areas of overlap, and areas which belong 'more or less' to a particular region or part of the world.

Another example of the same phenomenon of updating local terminology in gloss form while maintaining the general archaic nomenclature, which we have observed in Honorius's treatment of Germany, is found in the fifteenth-century *Das puch von der Forme und pildnuss der Welt* (ff. 85v–86r), which adds contemporary names almost as glosses on the original nomenclature for various regions. Examples include, for instance, in the chapter on Germany, additions to the sentence describing Suevia: 'Das ist Swaben landt' (f. 85v ll. 20–21).[42] Similarly for the Latin phrase 'Est in ea Noricus que et Bawaria, in qua est civitas Ratispona', the German adds glossing updating the two later names, resulting in 'Es ist auch darhin das landt Norica, daz yetzt hasst bauaria. Das ist Bayrnland. Darhin leyt dy stat Ratispana daz ist regenspurg' (f. 85v ll. 24–26).[43] (For the manuscript text, see the detail from the manuscript page reproduced in Figure 3.) *Bauaria* is thus glossed in the vernacular as *Bayrnland* and *Ratispana* as *Regenspurg*. The additions are not exclusive to Germany, and there are explanations, for instance, of the meaning of the Mediterranean, for which the Latin term

42 Cf. Flint, ed., *Imago mundi*, I.28 (p. 59).

43 Cf. Flint, ed., *Imago mundi*, p. 60.

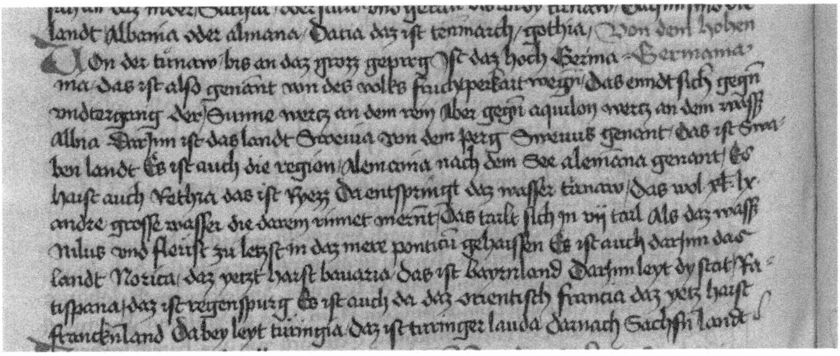

Figure 3: The chapter on Germany in Staatsbibliothek Preussischer Kulturbesitz Germ. Fol. 595, f. 85v detail

is used, glossed as 'the sea which is in the middle of the world' (f. 85v ll. 36–37). While no glosses of this kind are made for Spain, *Anglia* is glossed as *Engelland* (f. 86v l. 28).[44] It could be that the inconsistency in addition of these updated identifications in the vernacular has to do with the origins of the updates. While it could be part of the translation in original design, it could also be that these terms represent marginal glosses subsequently drawn into the translated main text.

Das puch von der Forme und pildnuss der Welt still awaits its own dedicated edition and study, but it can be said here on the basis of the brief overview that there is some updating even within this text, probably with the goal of helping the readers identify the places in question, which otherwise represents a faithful translation of the *Imago mundi*, maintaining its archaic flair well into the fifteenth century.

Overlapping Maps: The Cases of *paludes Meotides*, Spain, and Rome

Variation in the use of archaic and updated place-names in the *Imago mundi* and its adaptations glimpsed in the discussion above can be seen in terms of simultaneous coexistance of two conceptions of space, with two overlapping maps: the topographical and the cultural/political. One is timeless, and thus anachronism here is irrelevant, and the other is susceptible to updating. This theme is explored in this section by looking at three case studies: the *paludes Meotides*, Spain, and Rome, each of which areas is changed in adaptations in relation to Honorius's antiquated

44 Cf. Flint, ed., *Imago mundi*, I.28 and I.29 (p. 62).

verbal map. My argument is that these changes are enabled by the *Imago mundi*'s archaism.

Already in the seemingly straightforward description in *Imago mundi* I.21, which begins the section of the text dedicated to Europe, there is both archaism and ambiguity. The chapter sets the pattern for the descriptions of the regions of Europe in enumerating the first of the northern regions, the Ripheian Mountains (*Rifei montes*), the Don (*Tanais fluvius*), and the *Meotides paludes* (usually idenfitied as the Sea of Azov).[45] Since we will also be examining its vernacular rewritings further in *Transforming Europe*, I quote the whole chapter.

> Europa ab Europe rege, vel ab Europa filia Agenoris est nominata. In qua inprimis versus septentrionem sunt Rifei montes et Tanais fluvius, a Tanai rege dictus, et Meotides paludes, magno mari iuxta Theodosiam urbem seiungentes.[46]
> Europe is named after King Europs or after Europa, daughter of Agenor. In which first towards the north are Ripheian mountains and the River Don, named after King Tanaus, and the Maeotian Swamps, severing the Great Sea close to the City of Theodosia.[47]

45 Evelyn Edson identifies *Paludes Meotides* as the Sea of Azov; 'Maps in Context: Isidore, Orosius, and the Medieval Image of the World', in *Cartography in Antiquity and the Middle Ages: Fresh Perspectives, New Methods*, ed. by Richard J. A. Talbert and Richard W. Unger (Leiden: Brill, 2008), pp. 21–236 (p. 222).

46 Flint, ed., *Imago mundi*, I.21 (p. 59).

47 My translation. The alternative etymology for Europe is unusual; see Michael Wintle, *The Image of Europe: Visualising Europe in Cartography and Iconography Throughout the Ages* (Cambridge: Cambridge University Press, 2009), pp. 102–52; see also discussion of Europs in Caroline D. Eckhardt, 'One Third of the Earth? Europe Seen and Unseen in the Middle English Chronicles of the Fourteenth Century', *Comparative Literature*, 58.4 'The Idea of Europe' (2006), 313–38 (pp. 319–20). The Europs identification is present in other texts of the *Imago mundi* tradition, e.g. Connochie-Bourgne, ed., *Image du monde*, II.1, ll. 2131–34 (p. 818). The resulting indeterminacy of two possible origins for the name may thus be considered a feature of the *Imago mundi* tradition. This type of multiplicity or ambiguity might be argued to be part and parcel of the text's flexibility and adaptability, which we will revisit in later chapters. For further references to this legend, see Klaus Oschema, *Bilder von Europa im Mittelalter*, Mittelalter-Forschungen, 43 (Ostfildern: Thornbecke, 2013), pp. 165, 204–06. King Europs is mentioned by Frechulf of Lisieux (s. IX), for instance, and Frutolf of Michelbserg (d. 1103), both of which predate the *Imago mundi*, as well as by Otto of Freising (1114–1158) and Rudolf of Ems (1200–1254) among others who used the *Imago mundi*. Europs also appears in the Medieval Irish 'Book of Leccan'; see Hildegard Tristram, ed., *Sex aetates mundi: die Weltzeitalter bei den Angelsachsen und den Iren. Untersuchungen und Texte*, Anglistische Forschungen, 165 (Heidelberg, 1985), p. 104 n. 119, referred to by Oschema (p. 204 n. 26). See also Eckhardt, 'One Third of the Earth?', pp. 318–20; Doberentz, 'Die Erd- und Völkerkunde', pp. 298–301, 387–454; and the continuation of Doberentz's

Although commonly this chapter is referred to in discussions of medieval geography as delineating the boundaries of Europe, in this passage, these are not explicitly described as boundaries.[48] The position of the *paludes Meotides* between two parts of the sea, which seems to be implied here, seems to go against Evelyn Edson's identification of these 'swamps' as the Sea of Azov, but it may also be the result of Honorius's misunderstanding.[49] The Azov Sea and the Black Sea are connected by the Kerch Strait; the modern Feodosia (corresponding in its location to the Ancient Greek Θεοδοσία) is located on the Crimean Peninsula, on the Black Sea coast, south-west of the Kerch Strait (Figure 4). Honorius may be referring to the Kerch Strait as the *paludes Meotides*. These *paludes*, whatever they might be in Honorius's view, derive from archaic classical geography, and that matches his use of the name Theodosia, which, unlike the swamps, is identifiable.

The reference to Theodosia in the passage exemplifies the anachronism of *Imago mundi* geography, as the city was known as Caffa/Kaffa for much of the Middle Ages.[50] Indeed, this reference is dropped in several of the adaptations, including *Delw y Byd* Version A, the more extensive of the two Welsh translations, and in the fourteenth-century Anglo-Norman

article in *Zeitschrift für deutsche Philologie*, 13 (1882), 29–56; Karl Schorbach, *Studien über das deutsche Volksbuch Lucidarius und seine Bearbeitungen in fremden Sprachen*, Quellen und Forschungen zur Sprach- und Kulturgeschichte der germanischen Völker, 74 (Strassburg: Karl J. Trübner, 1894), p. 163. Europs does not feature in the context of an etymology for Europe in Honorius's source, Isidore's *Etymologies* (XIV.iv.1). However, Europs appears in the *Etymologies* in a different context at at XIV.iii.41.

48 The Eurasian boundary has been the subject of much discussion; W. H. Parker, 'Europe: How Far?', *The Geographical Journal*, 126 (1960), 278–97; Anthony Pagden, 'Europe: Conceptualising a Continent', in *The Idea of Europe: From Antiquity to the European Union*, ed. by Anthony Pagden (Cambridge: Cambridge University Press, 2002), pp. 33–54 (pp. 46–47); Wintle, *The Image of Europe*, pp. 9–10, 38, 41, 43, 51–52; Oschema, *Bilder von Europa*, pp. 211–12; and Leonid S. Chekin, *Northern Eurasia in Medieval Cartography: Inventory, Text, Translation, and Commentary* (Turnhout: Brepols, 2006). Already, Herodotus had questioned where the boundary lay; James S. Romm, 'Continents, Climates, and Cultures: Greek Theories of Global Structure', in *Geography and Ethnography: Perceptions of the World in Pre-Modern Societies*, ed. by Kurt A. Raaflaub and Richard J. A. Talbert (Chichester: Wiley–Blackwell, 2010), pp. 215–35 (pp. 215, 225); Wintle, *The Image of Europe*, p. 85. Some of the adaptations of *Imago mundi*, such as the *Image du monde*, give no Europe–Asia boundary; Chantal Connochie-Bourgne, 'Limites et diversités de l'Europe: le parti pris par Gossouin de Metz (*Image du monde*, 1245)', *De la Chrétienté à l'Europe : Actes du Colloque Orléans, mai 1993*, ed. by Bernard Ribémont (Orleans: Paradigme, 1995), pp. 49–62 (p. 59 n. 6). My point here is that this is already a feature of the Latin.

49 Edson, 'Maps in Context', p. 222.

50 For a discussion, see Michel Balard, 'The Greeks of Crimea under Genoese Rule in the XIVth and XVth Centuries', *Dumbarton Oaks Papers*, 49 (1995), 23–32.

Figure 4: *Imago mundi* locations mentioned in Chapter 3

Divisiones mundi (which, however, abridges the text considerably throughout and cannot be brought to the problem as a diagnostic witness insofar as omissions are concerned, since in this text those are endemic).[51] Generally, Version A of *Delw y Byd* is considered to be closer to the original Latin, but in retaining the Theodosia reference, it is the shorter Version B which is closer to the Latin text in this instance.[52] In the case of the Welsh translations,

51 See O. H. Prior, ed., *Divisiones Mundi*, in *Cambridge Anglo-Norman Texts* (Cambridge: Cambridge University Press, 1924), pp. 59–60. On the *Divisones mundi*, see also the discussion in Natalia I. Petrovskaia, 'Europe and the Holy Land in the British Branch of the *Imago mundi* Tradition', in *Crusading Culture and the Holy Land in Medieval Britain*, ed. by Kathryn Hurlock and Laura Whatley (Turnhout: Brepols, 2022), pp. 41–66. For more on the term *divisiones* see this book, pp. 106–09.

52 Henry Lewis and P. Diverres, ed., *Delw y Byd* (*Imago Mundi*) (Cardiff: Cardiff University Press, 1928), pp. 39, 93; and Natalia I. Petrovskaia, ed., *Delw y Byd. A Medieval Welsh Encyclopedia, MHRA Library of Medieval Welsh Literature* (London: MHRA, 2020), p. 53.

Version B does tend to maintain the form of the Latin place-names (going so far as to retain the Latin case endings on occasion), so this may well be an instance where the translators kept the references as they were because these were unfamiliar to them.[53]

Thus, for some of the *Imago mundi* rewritings archaism remains a feature. The *Semeiança del mundo* which, like *Delw y Byd*, is a very close rendering of *Imago mundi*, also displays this feature prominently. For example, the chapter on Spain which largely corresponds to the Latin, and also ends on the enumeration of the six provinces of Tarragona (Catalonia), Cartagena (Murcia), *Lusitania* (Roman province encompassing Portugal and neighbouring parts of modern Spain), Galicia, *Betica* (Roman province, corresponding roughly to modern Andalucía), and *Tinguitania*.[54] An important point to note in relation to this list of regions is that, as pointed out in the identifications provided in brackets, the nomenclature used is largely that of the Roman provinces. The text, therefore, demonstrates the archaising feature observed in much of the *Imago mundi* tradition.[55] A similar archaising tendency has been observed by Leszek Wojciechowski in his analysis of Isidore's description of Spain in the *Etymologies*.[56] As Wojciechowski points out, Isidore's description formed the base of Honorius's, but the latter had changed the emphases slightly for a new audience, providing only 'enough for a reader (receiver) to be able to place the country in the overall picture of the inhabited globe drawn (from the Isidorian perspective) by Honorius'.[57] In this respect, the *Semeiança* follows in that tradition. As Alan Deyermond observes, 'the description of the world offered by the *Semeiança* is not, as one might suppose from its date, a product of the XIIc Renaissance, but a late presentation of largely classical geographical beliefs'.[58] That the *Semeiança*, in the passage on Spain, does not update the text, is interesting in the light of other changes made in the translation

53 For the Welsh translators' attitudes towards the text, see Petrovskaia, ed., *Delw y Byd*, pp. 13–14.

54 A marginal gloss identifies Lusitania as the region *in qua est Lisebona* ('in which Lisbon is') in at least one manuscript of the *Imago mundi*: Cambridge, Corpus Christi College MS 66. For Anglo-Norman involvement in Lisbon, see Lucas Villegas-Aristizabal, 'Norman and Anglo-Norman Participation in the Iberian Reconquista *c.* 1018–*c.*1248' (unpublished PhD thesis, University of Nottingham 2007), pp. 160, 166–68, 176–85.

55 See this book, pp. 87–94.

56 Wojciechowski, 'Geographical Descriptions', p. 39.

57 Wojciechowski, 'Geographical Descriptions', p. 48.

58 Alan Deyermond, 'Building a World: Geography and Cosmology in Castilian Literature of the Early Thirteenth Century', *Canadian Review of Comparative Literature*, 23 (1996), 141–59 (p. 144).

(including the omission of the description of Britain, which is moved to the section on islands).[59]

It is helpful to think of these terminological shifts also as the product of the overlay of several different maps (such as is created by an artist's reconstruction on a photograph of the *Fori Romani* in a tourist booklet). There is a geographical space, which Honorius takes from Isidore, and it is overlaid by the political space, which, as we shall see in the discussion of the adaptations of the material in Gervase of Tilbury and Brunetto Latini below, is susceptible to replacement by a different overlaying map, albeit one which keeps the underlying geographical framework intact. Whether the overlaying map has sharply drawn boundaries does not affect the nature of the underlying geographical framework. The etymological and historical information given by the Latin text, just like its depiction of the world, is paradoxically both concrete and fluid, facilitating adjustment of the map, where necessary, by its translators and adaptators. We will now consider an example of this in the ecclesiastical rewriting of Rome by Gervase and Brunetto.

The *Imago mundi* was used by Gervase of Tilbury in the geograpical description in *Otia imperialia* (*c.* 1209–1215), an instructional work in the 'mirror of the princes' genre composed for the future Otto IV.[60] Gervase relied heavily on Honorius's text in his description of the world in the *Secunda decisio* (Book II), particularly in his description of Europe.[61] My discussion will focus on aspects of this description. While an extensive analysis of the extent of the use of *Imago mundi* in the *Otia imperialia* is not possible within the limits of the present study, it is worth taking the two texts side by side. Their production, though separated in time by almost a century, also shares some similarities of cultural context (possible in Honorius's case and fairly certain in that of Gervase) – partly English and

59 We return to the discussion of the category of Islands in Chapter 4.

60 Flint, ed., *Imago mundi*, pp. 11–12; O. Doberentz, 'Die Erd- und Völkerkunde in de Weltchronik des Rudolf von Hohen-Ems', *Zeitschrift für deutsche Philologie*, 12 (1881), 387–454 (pp. 412–18); Fortunata Latella, 'Come lavorava un intelettuale laico del medioevo: Gli *Otia imperialia* di Gervasio di Tilbury tra *intenvio* e *compilatio*', *Revista de literatura medieval*, 25 (2013), 103–34 (pp. 109, 113, 116); Jerzy Strzelcsyk, 'Gervase of Tilbury (*c.* 1165–*c.* 1234)', in *Trade, Travel, and Exploration in the Middle Ages. An Encyclopedia*, ed. by John Block Friedman and Kristen Mossler Figg, with Scott D. Westrem and Gregory G. Guzman (New York and London: Routledge, 2000), pp. 227–28 (p. 228); Wojciechowski, 'Geographical Descriptions', p. 49. For the date, see Gervase of Tilbury, *Otia Imperialia: Recreation for an Emperor*, ed. and trans. by S. E. Banks and J. W. Binns (Oxford: Clarendon Press, 2002), p. xxxix.

61 Strzelcsyk, 'Gervase of Tilbury', p. 228. See also the notes in Banks and Binns's edition, identifying the *Imago mundi* as the source of multiple passages throughout the text.

partly German. Probably an Englishman by birth, as identified by his name of Tilbury in Essex, Gervase travelled widely, having studied in Bologna, and had served Henry II, Archbishop William of Rheims, and William II of Sicily, before he finally settled in Arles, in Provence, from around 1190.[62] His career is even more transnational than that of Honorius and exemplifies the interconnected, networked world of medieval Europe that bred the broad interlingual family of *Imago mundi*–derived texts.

Like the *Imago mundi*, Gervase's text moves from the east to the west. His attention is more than that of Honorius, however, on history, and on matters ecclesiastical. He dedicates considerably more space to Rome than does Honorius, including additional information related to the Basilica of St Peter and references to the popes, not found in the former.[63] This is particularly striking when Gervase's text is taken alongside Latini's, composed a little later in the same century and also reliant on the geographical description of the *Imago mundi*. Latini, whose career was as international as that of Gervase, is much more concerned with the present, but he shares with Gervase an interest in ecclesiastical matters, barely visible in Honorius, as noted above.[64] Although Gervase dedicates much more space to the description of Rome than does Latini, they do have at least one addition in common, which is illustrative of the point being made here.

The addition concerns the description of the sees of the cardinal-bishops subject to the see of Rome. In Gervase's text, this reads: 'Roma papalis sedes est, quinque habens ecclesias patriarchales; suntque episcopatus cardinales et speciales eius isti: Ostiensis, Portuensis, Albanensis, Prenestinus, Sabinensis, Tusculanus, Tiburtinus, Reatinus' ('The papal see is in Rome. It has five patriarchal churches, and the following sees, whose bishops are cardinals, are subject to it alone: Ostia, Porto, Albano, Palestrina, Sabina, Tusculum, Tivoli, and Rieti').[65] In Brunetto Latini's text, the same passage reads: 'Et sachiés que li apostoles de Rome a desous lui .vi. evesques ki sont chardenals, celui d'Ostie et de Albani et de Portes et de Savine et de Tosquelain et de Penestraine' ('And know that the apostle of Rome has beneath him 6 bishops who are cardinals: that of Ostia and of Albano and of Porto and of Sabina and of Tusculum and of Palestrina').[66] Brunetto's text omits Tivoli and Rieti. The other sees on the list, though in a slightly different

62 Gervase of Tilbury, *Otia imperialia*, ed. and trans. Banks and Binns, pp. xxv–xxx, xxxviii.

63 Gervase of Tilbury, *Otia imperialia*, II.8, ed. and trans. Banks and Binns, pp. 260–71.

64 Jostkleigrewe, 'L'espace', pp. 373–74.

65 Gervase of Tilbury, *Otia imperialia*, II.8, ed. and trans. Banks and Binns, pp. 270–71.

66 Brunetto Latini, *Li Livres dou Tresor*, ed. by Francis Carmody (Berkeley: University of California Press, 1948), p. 116; my translation.

order, are the same. The editors of *Otia imperialia* identify the source of the passage as the *Provinciale Romanum*, and note that Rieti does not appear in that text at that point.[67] It seems that both Gervase and Brunetto use the same source to add to the information derived from *Imago mundi* here, but with minor alterations. Could it be that both relied on an intermediate source that already combined this with *Imago mundi*? The question will have to remain unanswered for now, but some tentative conclusions might be drawn based on this comparison.

This type of reference represents what is, perhaps, the most striking point of divergeance between the geography of Europe in the Latin original and in its rewritings, such as the *Trésor* and *Otia imperialia*.[68] The former provides the framework and basic information to which new material is added in the rewritings, given new meaning and a different emphasis. In the texts shown here the emphasis is religious. This might seem to support Hay's thesis, mentioned in the Introduction, that Christianity or Christendom was a much more important term for the period than Europe was.[69] Yet we are really back to the simile of the *Fori Romani* tourist booklet: with a Christian, ecclesiastical, framework superimposed in these texts upon the geographical structure provided by Honorius.[70] The underlying description of Europe which provides the framework does not carry the ecclesiastical connotations prior to the rewritings. These are added by later authors based on the needs and requirements of their new compositions, which are no longer narrowly geographical in focus.

To illustrate this point, it is worth considering a broader section of Latini's text focusing specifically on comparing his use of nomenclature with that of Honorius. We have seen that the Latin text uses the term *provincia* in its classical sense extensively. By contrast, though in his discussion of Italy Latini writes of its *provinces*, it becomes clear almost immediately that these are ecclesiastical rather than secular, as the section begins with an enumeration

67 Gervase of Tilbury, *Otia imperialia*, ed. and trans. Banks and Binns, p. 271 n. 64. For the text, see *Calendarium Historico-Christianum*, ed. by A. J. Weidenbach (Regensburg, 1855), pp. 264–84, at 264. For more on the *Provinciale*, its circulation and use by Gervase, see Fabrice Delivré, 'Du nouveau sur la "Liste de Florence": La chronique du Pseudo-Godel (v. 1175) et la prehistoire du Provinciale Romanum du XIIIe siècle', *Bibliothèque de l'École des chartes*, 167 (2009), 353–74, esp. p. 366 n. 64; and Benedict Wiedemann, 'The Joy of Lists: The *Provinciale Romanum*, Tribute and *Ad limina* Visitation to Rome', *Revue d'histoire ecclésiastique*, 115 (2020), 61–97.

68 For further discussion of Latini's *Trésor* and its interest in church territories, see pp. 151–53.

69 See p. 20.

70 As Jostkleigrewe points out, the ecclesiastical re-focusing of the text by Latini also constitutes an updating of the archaic Roman imperial geographical framework used by Honorius; Jostkleigrewe, 'L'espace', p. 374.

of the cardinals subject to the Pope in the the *Provinciale Romanum*–derived passage discussed above.[71] Latini's entire verbal map of Italy is therefore a map of the Italian church: his focus is on the archbishoprics and bishoprics exclusively, with occasional reference to cities. A typical example is his description of Lombardy: 'Après i est Lombardie, ou est Boloigne la crasse et .iii. autres cités, et l'archeveschié de Melan, ki dure dusc'a la mer de Gene, et la cités de Saone et de Albinge, et puis jusc'a la terre de Ferrere, ou il a .xviii. evexchiés' ('After this is Lombardy, in which is located Bologna the fertile and many bishoprics, and the archbishopric of Mila, which extends to the Ligurian Sea and the city of Savona and Albenga and then up to the land of Ferrara, where there are 18 bishoprics').[72] There is no trace of secular power structures. The same pattern can be traced throughout Latini's description of Europe, and I will not belabour that point here.

As with Honorius, there is contrast in Latini's useage between terminology used in the Europe section on the one hand, and in the descriptions of Asia and Africa, on the other, but it is even more striking. If in Honorius, as we have seen, differences were limited to absence of presence of *provincia* references, based on the extent of the Roman Empire, there is a more complex pattern of differentiation in Latini.[73] In his description of Asia, he uses the terms *païs*, *province*, *terre*, and refers to *cités* and *villes*.[74] Within the description of the region of Asia, the only reference to ecclesiastical territories comes in the section beginning with the description of Jerusalem: 'En celui païs a mains patriarches et archevesques et evesques; selonc l'etablissement de sainte eglise, ki sont par conte .c. et xxxiii' ('In that country there are many patriarchs, archbishops and bishops, according to the establishment of Holy Church, and these total 133').[75] Although we cannot expect to find references to bishoprics where none existed, the contrast between the ecclesiastical focus of the description of Europe and the complete lack of focus on religion in the discussion of Asia (outside the Holy Land) and Africa, is striking. Conversely, while in the section on Europe, Latini uses the terms *païs* and *terre*, the term *province* (which we have seen in Honorius's text) is conspicuous by its absence. The marked

71 See above, pp. 100–01.

72 Brunetto Latini, *Li Livres dou Tresor*, ed. by Carmody, p. 116; *The Book of Treasure* (*Li Livres dou Tresor*), trans. by Paul Barrette and Spurgeon Baldwin, Garland Library of Medieval Literature Series B, vol. 90 (New York and London: Garland, 1993), p. 93.

73 See the discussion of Honorius's provincial nomenclature above, pp. 87–93.

74 See, for instance, Brunetto Latini, *Tresor*, ed. by Carmody, p. 112.

75 Section 27; Brunetto Latini, *Tresor*, ed. by Carmody, p. 114; trans. by Barrette and Baldwin, *The Book of Treasure*, p. 91.

change in the use of terminology when it comes to Europe is strikingly similar to the trend observed in the *Imago mundi*. Whereas the pattern in that text is, as we have seen, based on the political jurisdictional structures of the Roman Empire, that in the *Trésor* is very clearly dictated by the extent of the political jurisdictional structure of the Roman Catholic Church (i.e. bishoprics and archbishoprics). Ultimately, of course, it had been the Roman provinces that had formed the basis of ecclesiastical territorial structures, and thus Latini's rewriting of Honorius parallels historical developments.[76]

In his geography, Latini's main interest in ecclesiastical jurisdictions is almost to the exclusion of other interests. Though the use of *province* is not exclusive to the regions beyond Europe in Latini's text where the Church does not hold jurisdiction (in Asia and Africa) forcing him to forgo ecclesiastical structures, he defaults to the *Imago mundi* description of secular regions, and his use of the term *province* for Thebaid and Bythinia, as we have seen, represents an anachronistic reference to the Roman Empire (increasingly so in Brunetto Latini's text).[77] The focus of the geographical description on ecclesiastical rather than secular territorial units might be related to the unstable nature of the latter in the mid-thirteenth century, as a result of the struggle between the papacy and the empire.[78]

The ecclesiastical framework seen here, while prominent, is dependent on the geographical structure provided by the *Imago mundi* as base text and cannot function without it. In rewriting this material Gervase and Brunetto do not so much transform Europe as they overlay another layer, an ecclesiastical layer, on top of the geographical material provided by Honorius. This is a purposefully scholarly exercise, for as Simon Gaunt points out, the *Trésor*, like the the *Image du monde* to which he compares it, is governed by 'a preoccupation with Christian geography' and explicitly relies on the authority of written sources, rather than on first-hand

76 A clear brief overview, in context of the use of the term 'province', is given in Peter J. Heather, *The Restoration of Rome: Barbarian Popes and Imperial Pretenders* (Oxford: Oxford University Press, 2013), p. 305; for a more detailed description of the process in a Gaulish context, see Yitzhak Hen, 'The Church in Sixth-Century Gaul', in *A Companion to Gregory of Tours*, ed. by Alexander C. Murray (Leiden and Boston: Brill, 2015), pp. 232–55 (pp. 238–40).

77 See discussion on p. 90.

78 A brief account of this struggle is given in the *Trésor*, I.lxxxviii; Brunetto Latini, *Li Livres dou Tresor*, ed. Carmody, pp. 80–81. For discussions see, for instance, David Abulafia, *Frederick II: A Medieval Emperor* (Oxford: Oxford University Press, 1988), pp. 63–66; and Brett Edward Whalen, *The Two Powers: The Papacy, the Empire, and the Struggle for Sovereignty in the Thirteenth Century* (Philadelphia: University of Pennsylvania Press, 2019), pp. 96–121.

experience.[79] This constitutes not only an act of self-positioning within the encyclopedic tradition, explicit in both authors' works, but the presentation of the work as a link within the *translatio* process of information derived from *auctoritas* ('authority').[80] These works build their authority on that of their model, and the changes they make, and the points they make, are facilitated by Honorius. Though the ecclesiastical framing in the *Trésor* and *Otia imperialia* seemingly supports Hay's thesis (that the precursor of the concept of 'Europe' in the Middle Ages was 'Christendom'), they can only function in counterbalance to the pre-existing map, which is archaic and – a point we turn to in Chapter 4 – fuzzy.[81] In Chapter 4 I argue that when we imitate Ruggiero on the hippogriff in attempting to 'leave Europe' in the *Imago mundi* texts, we cross no sharply delineated boundary.

79 Gaunt, *Marco Polo's 'Le Devisement'*, pp. 70, 71. For an example of the increased focus of Brunetto's work on ecclesiastical geography, compared to Honorius's text, see above, p. 100.
80 Simon Gaunt observes that both texts are 'a tissue of translated quotation', citing the reference to the act of translation in the opening of *Image du monde* and of the *Trésor*; Gaunt, *Marco Polo's 'Devisement'*, p. 71 and esp. n. 66.
81 For the use of the term 'Europe' in chronicle writing, see in Caroline D. Eckhardt, 'One Third of the Earth?', 313–38.

4. *Space*. Geographical Regions as Fuzzy Sets

Abstract: This chapter addresses the terminology relating to boundaries, used in describing the three parts of the world in the *Imago mundi*, *Image du monde*, and Brunetto Latini's *Trésor*. This shows that contrary to the conventions adopted in modern discussions, the geographical description in these texts is predominantly cumulative rather than divisive. Further, the areas delimiting the three parts of the world are shown to have been conceived of as zones rather than lines, indicating that both the areas so defined and the boundary zones themselves should be seen as fuzzy. This opens up a new solution to the problem of the re-assignation of some areas to other parts of the world in the *Image du monde*.

Keywords: fuzzy sets; borders; boundaries; Mediterranean; Europe and Africa; *Image du monde*

As we have seen in Chapter 2, in *Imago mundi* I.7 only the Mediterranean is mentioned as a separation between the three parts of the world.[1] The discussion here will investigate this further. This chapter therefore begins by examining the 'divisive' terminology in *Imago mundi* and its adaptations. I propose to show that what might at first glance appear as borders frequently fit more easily into the concept of 'area of overlap' (or frontier in its spatial sence as area rather than line).[2] We will then look at how regions

1 Quoted in Chapter 2, p. 46.

2 For suggestions that the medieval tripartite scheme was not necessarily divisive, see Benjamin Braude, 'The Sons of Noah and the Construction of Ethnic and Geographical Identities in the Medieval and Early Modern Periods', *The William and Mary Quarterly*, 54 (1997), 103–42 (p. 109); David Woodward, 'Medieval *Mappaemundi*' in *History of Cartography I: Cartography in Prehistoric, Ancient, and Medieval Europe and the Mediterranean*, ed. by J. B. Harley and David Woodward (Chicago and London: University of Chicago Press, 1987), pp. 286–370 (p. 290); Caroline D. Eckhardt, 'One Third of the Earth? Europe Seen and Unseen in the Middle English Chronicles

Petrovskaia, N.I. *Transforming Europe in the Images of the World, 1110–1500. Fuzzy Geographies*. Taylor & Francis Group, 2025
DOI 10.5117/9789048563166_CH04

of the world in the *Imago mundi* can be read as fuzzy sets, before revisiting the issue of problematic boundaries by looking in detail at the apparent re-assignation of apparently European regions to Africa and Asia in the *Image du monde*. In this discussion I will show how the fuzzy set model combines with previous suggestions, the historico-geographical model in particular, to help us understand this phenomenon. Understanding the three parts of the world as partially overlapping fuzzy sets also opens up the way to reading the Mediterranean itself as a fuzzy set. The concluding section of this chapter will accordingly show how this reading helps us understand the cumulative structure of the world in the *Imago mundi*, as well as the changes which its discussion of islands undergo in the various vernacular rewritings. The fact that these changes are often brought about by the reordering of the the itinerary will lead us to our next chapter, and the discussion of the hodoeporical nature of the text's design.

We therefore now continue our flight on the hippogriff, paying attention to whether we encounter any borderlines as we approach the regions where Europe ends for Ruggiero as he sees the pillars that Hercules set up as a sign.

Divide or Describe? *Dirimere, deviser*

The term used in *Imago mundi* to refer to the role of the Mediterranean in the spatial layout of the three parts of the world is *dirimitur*, the present passive indicative of the verb *dirimere/dirimo* 'to take apart, separate, divide'.[3] The lack of reference to the rivers Don and Nile, conventionally added as boundaries between Europe and Asia and Africa and Asia respectively in descriptions of medieval geography (usually based on the *mappae mundi*), is noteworthy. Both rivers are mentioned elsewhere in the text, and it might also be argued, tentatively, that by limiting the reference to the Mediterranean here the emphasis in the sentence is placed on the regions themselves rather than the boundaries between them.[4] The Mediterranean thus appears to be described as a boundary. It is important to note that its

of the Fourteenth Century', *Comparative Literature*, 58.4 'The Idea of Europe' (2006), 313–38 (p. 324).

3 Honorius Augustodunensis, *Imago mundi*, ed. by Valerie I. J. Flint, *Archives d'histoire doctrinale et littéraire du Moyen Âge*, 49 (1982), p. 52; Charlton T. Lewis and Charles Short, *A Latin Dictionary* (Oxford: Clarendon Press, 1879), s.v. *dirimo*.

4 A similar trend highlighting centres such as cities over borders and frontiers was noted in medieval Muslim geography by Ralph W. Brauer in *Boundaries and Frontiers in Medieval Muslim Geography* (Philadelphia: American Philosophical Society, 1995).

function separating the other regions does not seem to be at odds with its being an area of space. We are dealing at the very least with a frontier as zone or area, and not as a line.[5] We will come back to this issue when looking at the Mediterranean as a fuzzy set later on in this chapter.

The word *dirimere* is translated in the French text as *deviser*, which means 'divide' (but not exclusively, as shall be seen below). Let us focus on how the word *deviser* is employed in the section translating *Imago mundi* in the *Image du monde*, the thirteenth-century *livre du clergie* whose Book II corresponds to *Imago mundi* I. As with the Latin text, in the *Image du monde* the apparent divisive structure – if it is indeed divisive in this case – is limited to the initial description of the three regions by the use of the term *deviser*. The rest of the description neatly fits the definition of sets: 'De ces .III. parz qui sont nommees / Tient chascune maintes contrees / Et maintes riches regïons / Dont nous dirons auques les nons' ('Of these three parts which are named / each has many countries / and many rich regions / of which we will give some names').[6] After France, which is the last region of Europe mentioned in the text, Europe includes 'toute la terre / Ki est jusqu'a mons de Mongieu' ('all the earth / which is as far as the mountain of Mongieu'), that is as far as the Alps.[7]

The *Image du monde* presents the tripartite composition of the inhabited world as *Si est en .III. parz devisez*.[8] Though this appears to be a literal translation of the Latin, the phrase nevertheless poses an interpretative difficulty, as in medieval French, *deviser* has three meanings: 'describe', and 'organise' or 'put in order' as well as 'divide'.[9] Elsewhere in the text,

5 A rather later geographical text, the sixteenth-century *Cosmographia* of Peter Apian/Petrus Apianus, uses *dirimitur* in relation to a region (*regio*) and an ocean: 'A parte meridiei Oceanus Meridionalis vocatur, quo hac ipsa regio dirimitur ab alia quadam nondum lustrata terra'; *Cosmographia, siue Descriptio universi orbis…* (Antwerp: Jan Bellerus, 1584), p. 175. The work was originally published in 1524; for a discussion, see Margaret Gaida, 'Reading *Cosmographia*: Peter Apian's Book-Instrument Hybrid and the Rise of the Mathematical Amateur in the Sixteenth Century', *Early Science and Medicine*, 21 (2016), 277–302.

6 Gossouin de Metz, *L'Image du monde, une encyclopédie du XIIIe siècle: edition critique et commentaire de la première version*, ed. by Chantal Connochie-Bourgne (unpublished PhD thesis, University of Paris, Sorbonne, Paris IV, 1999), p. 818, ll. 2143–46; my translation.

7 *Image du monde*, ed. Connochie-Bourgne, p. 840, ll. 3000–01. Mongieu derives from the Latin Mons Iovis and refers to what is now known as St Bernad's Pass in the Alps; see, for instance, Claude Lecouteux, 'Aspects mythiques de la montagne au Moyen Age', *Le Monde Alpin et Rhodanien: Révue régionale d'ethnologie*, 1.4 (1982), 43–54 (p. 45) ; available at http://www.persee.fr/doc/mar_0758-4431_1982_num_10_1_1140 (accessed 11 April 2024).

8 Connochie-Bourgne, ed., *Image du monde*, p. 817, l. 2118.

9 *Dictionnaire du Moyen Français* (*DMF*), s.v. *deviser* ; available at http://www.atilf.fr/dmf/definition/deviser (accessed 11 April 2024). See also the discussion in Simon Gaunt, *Marco Polo's*

deviser is used in the sense of 'describe' rather than 'divide' and it is possible that here the sense is also 'organise', or even, albeit with less likelihood, 'describe' (referring to the text's own activity in describing the three regions). The Europe section in the French text commences as follows: 'Puis qu'Asie devisee avons, / Europe après deviserons' ('Now that we have described Asia / After we will describe Europe)'.[10] Here, the verb *deviser* is being used in the sense of 'describe'. Thus, in the case of tripartite 'division' also Gossouin may be composing the inhabited world from three sections rather than dividing it. We are faced here, therefore, in the first instance with the fuzziness not of the boundaries of Europe but rather of the terminology itself used to describe the region. The important points here are that the regions are described as sets and that the term used, *deviser*, does not necessarily mean 'divide' although it can have that meaning.

Brunetto Latini also uses the same term in his *Trésor*. He introduces Europe in a passage reminiscent of *Imago mundi* I.21, but that, unlike the Latin text, explicitly refers to the tripartite structure in terms of division and separation by using a preposition *de* alongside the fuzzy term: 'Europe est une partie de la terre ki est devisee de celui d'Asie la u est li estrois dou Bras St. George et es parties de Constantinoble et de Grece…'[11] ('Europe is a part of the earth which is separated from that part of Asia where the Hellespont is, in the areas of Constantinople and Greece').[12] Note that while it does appear that here we are dealing with division rather than description, the entity that is doing the dividing is not necessarily a line. Barrette and Baldwin render *et es parties* as 'in the areas' but technically the text has the coordinating conjunction (*et*), which might imply that the boundary here is fuzzy. If we interpret the text to mean 'that part of Asia where the Hellespont is, and the areas of Constantinople and Greece', the Hellespont no longer constitutes a boundary line between Asia and Europe. Rather, the general vicinity of the Hellespont, Constantinople, and Greece is the area where Asia becomes Europe and Europe becomes Asia. These are marginal members, I would argue, of both fuzzy sets, and we are dealing

'Le Devisement du Monde': Narrative Voice, Language and Diversity (Cambridge: D. S. Brewer, 2013), pp. 150–51.

10 *Image du monde*, ed. Connochie-Bourgne, p. 839, ll. 2977–78; my translation.

11 Brunetto Latini, *Li Livres dou Tresor*, ed. by Francis Carmody (Berkeley: University of California Press, 1948), p. 115.

12 Brunetto Latini, *The Book of Treasure* (*Li Livres dou Tresor*), trans. by Paul Barrette and Spurgeon Baldwin, Garland Library of Medieval Literature Series B, vol. 90 (New York and London: Garland, 1993), p. 91.

with a frontier as area of overlap, not as line.[13] This point can be pressed further and has interesting implications for the history of the development of the medieval idea of borders.[14] The use of the word *parties* ('parts'/'areas') for the areas of Constantinople and Greece in Latini's text is significant, as will become clear from the discussion of his terminology further below. As much of the spatial terminology discussed in this chapter, the term is fuzzy. In its spatial sense, the word *partie* was used for the parts of the world (as in this passage for Europe), for a part of space or 'territory', 'region', or 'direction'.[15] Although 'direction' is unlikely to be a good translation for the passage quoted above, the fact that the term's semantic range includes the notion of trajectory is significant in the light of the hodoeporical nature of the description in the *Imago mundi* tradition which we will turn to in Chapter 5. It is worth carrying with us as we proceed the awareness of possible ambiguity in references to division, description, and definition in both the Latin text and in its Francophone translations discussed in this chapter.[16] If the *parties* are where the border lies, then that border is (in these texts at least) fuzzy – a frontier area of overlap between fuzzy sets.

Regions as Fuzzy Sets

This chapter began with a brief discussion of fuzzy boundaries in the *Imago mundi* tradition. We turn now to the notion of the regions themselves as fuzzy sets. The phenomenon is best observed in the *Imago mundi* chapters that mark the switches between sections of 'Asia'/'Europe', 'Europe'/'Africa' in turn. The quotation below presents the switching point from Asia to Europe,

13 For the notion of frontier as area, see 'Preface' to *Medieval Frontiers: Concepts and Practices*, ed. by David Abulafia and Nora Berend (New York and Abingdon: Routledge, 2016), pp. x–xvi, and above, p. 29–30. The idea of Greece and Constantinople as marginal members of the Europe *and* Asia fuzzy sets echoes the ambiguities of western European attitudes towards the Greeks/ Byzantines in the Middle Ages; see Jonathan S. C. Riley-Smith, *The Crusades: A Short History* (London: Athlone, 1987), pp. 2–3, 18–24, 90–91, 98–99; Malcolm Barber, 'Western Attitudes to Frankish Greece in the Thirteenth Century', *Mediterranean Historical Review*, 4 (1989), 111–28; Natalia I. Petrovskaia, *Medieval Welsh Perceptions of the Orient* (Turnhout: Brepols, 2015), pp. 33–35, 125–27.

14 For the current state of the scholarship, see Abulafia and Berend, eds, *Medieval Frontiers*.

15 *DMF*, s.v. *partie*; available at http://www.atilf.fr/dmf/definition/partie (accessed 26 April 2020).

16 Not all of the translations of *Imago mundi* are ambiguous in their terminology; the Welsh *Delwy Byd*, for instance, uses *rannu* ('divide'); Natalia I. Petrovskaia, ed., *Delwy Byd: A Medieval Welsh Encyclopedia*, MHRA Library of Medieval Welsh Literature (London: MHRA, 2020), p. 49; *Geiriadur Pryfisgol Cymru*, s.v. *rhannaf: rhannu*; available at https://geiriadur.ac.uk/gpc/gpc.html (consulted 17 September 2021).

at Chapters 20 and 21. The Latin text does not provide, in the chapters at the end and at the beginning of sections, any reference to particular defined linear boundaries. It merely lists the regions' components. These correspond to the definition of a set as an entity composed of a group of smaller entities. The boundaries between the regions would thus be boundaries between sets. The presence of overlap and of regions that belong to one or the other to different degrees would then constitute overlap between fuzzy sets (where a fuzzy set is one to which something might belong to a greater or lesser degree).

Taken in this context, the Riphaean mountains and the Don can be seen not as boundaries but rather as marginal/overlapping members of the fuzzy sets described. That the cumulative nature of the description in *Imago mundi* I.21 is not an anomaly can be aptly illustrated by examining the following short chapter, I.22 ('on Scythia').[17] This builds on the overall description of Europe given in the previous chapter: 'A Tanai fluvio est Scithia inferior, que versus meridiem usque ad Danubium porrigitur. In hac sunt iste provincie, Alania, Dacia, Gothia' '(From the River Don is Lower Scythia, which is extended towards the south as far as the Danube. In this are the following provinces: Alania, Dacia, Gothia').[18] Whilst at first glance, beginning the description in this chapter with the Don might suggest that it is used as a boundary, the reference to Europe in the chapter above, *in qua inprimis versus septentrionem sunt Rifei montes et Tanais fluvius*, suggests that it is perceived to lie within Europe, rather than necessarily merely delimiting its border.[19] The *Imago mundi* thus presents the region of Europe as what we can now describe as a fuzzy set, using the Don and the Mediterranean members of the set, without specifying these necessarily as borders, leading to a flexibility in the framework provided for later authors.

The same can be observed in the transitional chapters between Europe and Africa (I.29–30).[20] No boundary is provided by Honorius to separate the two parts of the world: rather, they are built up of components, meeting the formal definition of a set.[21] In this case the components are the regions

17 Flint, ed., *Imago mundi*, p. 59.

18 Ibid.

19 Ibid. For the idea of the Don as a boundary of Europe, inherited by the medieval T-O tradition from the Greeks and Romans, see, for instance, Michael Wintle, *The Image of Europe: Visualising Europe in Cartography and Iconography Throughout the Ages* (Cambridge: Cambridge University Press, 2009), pp. 41, 91.

20 Flint, ed., *Imago mundi*, pp. 62–63.

21 The description of Africa, for instance, beginning with the etymology, lists the various regions in the following manner: 'Huius prima provincia est Libia, a regina eiusdem nominis

named in the chapters dedicated to the sections on Europe, Asia, and Africa, respectively.

The same phenomenon has been observed by Caroline D. Eckhardt in her study of one of the world maps illustrating Ranulf Higden's *Polychronicon* where, she notes, 'Europe's constituent elements may be visible, but the perimeters are unmarked, and continental identities as such are effaced by other structural patterns'.[22] I suggest that in the case of the *Imago mundi* these 'structural patterns' can be usefully described as fuzzy sets. Taken in the light the existence of compositional geography in both cartographic and textual formats, where the regions constitute fuzzy sets, rather than being divided into sections by boundaries or frontiers, the existence of such maps alongside divisive T-O types appears less disruptive. Rather than constituting an exception, the *Polychronicon* maps and the *Imago mundi* text represent a shared tradition of compositional fuzzy geography.

This way of reading the material also makes it possible to resolve another problem, that posed by the *Image du monde* and its re-assignation of some European regions to Africa. A workable solution has already been proposed by Georg Jostkleigrewe, in the form of a 'historico-geographical' framework, but he thought it was a unique case. The discussion below will show that the fuzzy set model works well in combination with Jostkleigrewe's solution, and that it also helps show that this case – and his solution to it – have a broader relevance.

Overlap between Fuzzy Sets in the *Image du monde*

The *Image du monde* introduces one of the most fascinating changes to the description of Europe within the *Imago mundi* tradition.[23] It reassigns some regions from one of the three parts of the world to another: it attributes

dicta. [...] Inde est Cirenaica, [...] Hec et Pentapolis, [...] Berenice, Arsinoe, Ptolomaide, Apollonia, Cirene' ('Of which the first province is Lybia, named after a queen of that name. [...] Thence is Cyrenaica [...] This is Pentapolis, named after 5 cities, namely: Berenice, Arsinoë, Ptolemais, Apollonia, Cyrene'); Flint, ed., *Imago mundi*, p. 63 (my translation).

22 Eckhardt, 'One Third of the Earth?', p. 324 (p. 325 for the map itself). Eckhardt refers to the author as 'Ralf of Higden'. See also above, p. 29 n. 53. For the text of the *Polychronicon*, see Ranulf of Higden, *Polychronicon Ranulphi Higden Monachi Cestrensis; together with the English Traslation of John Trevisa and of an Unknown Writer of the Fifteenth Century*, ed. by Churchill Babington and Joseph R. Lunby, Rolls Series, 9 vols. (London: Her Majesty's Stationary Office, 1865–86).

23 There are multiple other, mainly minor, alterations, which are beyond the scope of the present study. Another major alteration, the relabelling of the whole of Asia as *India*, is examined in Natalia I. Petrovskaia, 'Mythologizing the Conceptual Landscape: Religion and History in

Jerusalem, Syria, Greece, a part of Italy, Spain, and Gascony to the region of Africa.[24] It also, in a contradictory or complementary movement, simultaneously assigns the first two also to Asia, and parts of Greece also to Europe.[25] As O. H. Prior observes, this reordering of the world structure confuses not only modern scholars but also the text's fifteenth-century English translator, William Caxton, but cannot be ascribed to a mere scribal error. In an early version of the French text, since it is present in all manuscripts of all of its versions.[26] Caxton's reaction to Gossouin's text, elegantly characterised by Prior as 'mild protest', suggests that whatever the reason behind the reordering of the geography in the French text may have been in the thirteenth century, by the fifteenth that reason was no longer obvious.[27] If we are willing to ascribe to medieval scribes a modicum of sense and to medieval readers a modicum of thought, we come inevitably to the conclusion that this must have made sense at some point to the scribes and readers of these manuscripts! On this principle, the alteration has received ample scholarly attention, in particular from Prior, Chantal Connochie-Bourgne, and Jostkleigrewe.[28] The intention of the following is to contribute to this

Imago Mundi, Image du Monde and *Delwy Byd*', in *Landscape and Myth*, ed. by Matthias Egeler, Medieval Texts and Cultures of Northern Europe (Turnhout: Brepols, 2019), pp. 195–211.

24 *Image du monde*, ed. Connochie-Bourgne, p. 840, ll. 3009–10.

25 Georg Jostkleigrewe, 'L'espace entre tradition et innovation: la géographie symbolique du monde et son adaptation par Gossouin de Metz', *Construction de l'espace au Moyen Age: Pratiques et représentations. XXXVIIe Congrès de la SHMES. Mulhouse, 2–4 juin 2006* (Paris: Publications de la Sorbonne, 2007), pp. 369–78 (p. 375), and further discussion below.

26 Caxton, *Mirrour of the World*, ed. by Oliver H. Prior, EETS, e.s. 110 (London: Kegan Paul, Trench, Trübner, 1913), p. xvi; for the multiple versions of the *Image du monde*, see the introduction to *Image du monde*, ed. Connochie-Bourgne, pp. 10, 27–30, 58–59; see Centili, 'La tradition manuscrite de l'*Image du monde*: Fortune et diffusion d'une encyclopédie du XIIIe siècle' (unpublished PhD dissertation, Paris, Ecole des Chartes, 2005). This section of the text was subject to one of the very few instances of intervention and comment by Caxton: 'And how be it that the Auctour of this book saye that thise contrees ben in Affryke, yet, as I vnderstonde alle thise ben within the lymytes and boundes of Europe'; Caxton, *Mirrour of the World*, ed. Prior, pp. 93–94; James A. Knapp, 'Translating for Print: Continuity and Change in Caxton's *Mirrour of the World*', *Disputatio*, 3 (1998): 64–90 (pp. 64, 73–74).

27 Caxton, *Mirrour of the World*, ed. Prior, p. xvi.

28 Caxton, *Mirrour of the World*, ed. Prior; Chantal Connochie-Bourgne, 'Le cas de l'*Image du monde*: une encyclopédie du XIIIe siècle, ses sources antiques, l'apport médiéval', in *La transmission des connaissances techniques: tables rondes Aix-en-Provence, avril 1993–mai 1994*, ed. by M.-Cl. Amouretti and G. Comet, Cahier d'histoire des techniques, 3 (Aix-en-Provence: Publications de l'Université de Provence, 1995), pp. 87–98; Chantal Connochie-Bourgne, 'Limites et diversités de l'Europe: le parti pris par Gossouin de Metz (*Image du monde*, 1245)', in *De la Chrétienté à l'Europe. Actes du Colloque Orléans, mai 1993*, ed. by Bernard Ribémont (Orleans: Paradigme, 1995), pp. 49–62 (pp. 52–55, 58); and Jostkleigrewe, 'L'espace', pp. 375–78. The issue is also discussed in Elisée Dion, 'Penser le monde en langue vulgaire et en images: enjeux sociaux

discussion by showing some of the results that can be attained by also examining this material through the prism of fuzzy set theory.

Prior's explanation for the phenomenon is that there was a distinct tradition which saw 'Africa as merely a province of Europe'.[29] Although Prior provides a number of examples, from Varro, Sallust, Orosius, Gervase of Tilbury, and Ranulph Higden's *Polychronicon*, the explanation is not altogether satisfactory, since this tradition is not represented by Gossouin's primary source for this section of the text: the *Imago mundi*.[30] Chantal Connochie-Bourgne, who revisited the issue as part of her work on the verse version of the *Image du monde*, suggests that the geographical reshuffling here can be explained by postulating that 'Europe' was seen as a cultural unit, and the lands reassigned by Gossouin to Africa in this text were regions that at some point had been subject to Islam.[31] Connochie-Bourgne's theory is that for Gossouin, the Alps represented a boundary for his Europe, limited to Western Christianity.[32]

Two arguments can be offered in favour of this interpretation. In the first place, the treatise as a whole self-identifies as a *livre de clergie*, signifying

et dimensions visuelles de la transmission du savoir dans l'Image du monde (XIIIe–XVe siècles)' (unpublished PhD dissertation, Université de Reims and Eberhard-Karls-Universität, 2018); I have not been able to consult this thesis but it is discussed in Christoph Mauntel, *Die Erdteile in der Weltordnung des Mittelalters: Asien – Europa – Afrika* (Stuttgart: Anton Hiersemann Verlag, 2023), pp. 253–54. I am grateful to Christoph Mauntel for this reference.

29 Caxton, *Mirrour of the World*, ed. Prior, p. xvi.

30 Ibid., n. 2.

31 Connochie-Bourgne, 'Le cas de l'*Image du monde*', pp. 91–93; Compare also Jerry Brotton's interpretation of the elision of the words *Europa* and *Asia* in the Berlinghieri's atlas (Florence, 1482); Brotton, *Trading Territories: Mapping the Early Modern World* (London: Reaktion Books, 1997), pp. 96–97, cited in Wintle, *Image of Europe*, p. 39. For a much more recent period, Eric Hobsbawm also mentions perceptions of Spain and South Italy as belonging more to Africa than Europe; Hobsbawm, 'The Curious History of Europe', in *On History* (London: Weidenfield & Nicholson, 1997), pp. 217–27 (p. 223).

32 Her interpretation of the world map of *Image du monde* thus seems to echo Denys Hay's influential idea of medieval 'Europe' as a cultural term equivalent to 'Christianity'; Connochie-Bourgne, 'Limites et diversités', p. 54. Connochie-Bourgne sees this as an echo contemporary discourse, particularly that of Jacques de Vitry, referring to what he saw as the Church's shrinking borders; Connochie-Bourgne, 'Limites et diversités', p. 56. Note that Jacques de Vitry may also have used the *Imago mundi*; Karl Schorbach, *Studien über das deutsche Volksbuch Lucidarius und seine Bearbeitungen in fremden Sprachen*, Quellen und Forschungen zur Sprach- und Kulturgeschichte der germanischen Völker, 74 (Strassburg: Karl J. Trübner, 1894), p. 162. See also William Chester Jordan's analysis of what would have been described as 'Christendom' territorially by contemporaries *c.* 1250 (which coincides largely with Gossouin's 'Europe'); Jordan, '"Europe" in the Middle Ages', in *The Idea of Europe: From Antiquity to the European Union*, ed. by Anthony Pagden (Cambridge: Cambridge University Press, 2002), pp. 72–90 (pp. 75–76).

that it is a vernacular text of encyclopedic nature of which the ostensible self-pronounced purpose is to improve the spiritual well-being of its lay audience.[33] A focus on matters religious is apparent throughout the text, as it relates the phenomena it treats to biblical events, to a greater degree than the *Imago mundi*, particularly in those Books of *Image du monde* not based on that Latin treatise.[34] We are not dealing here with an overtly ecclesiastical map along the lines of those by Latini or Gervase.[35]

The second point is contextual. The composition of *Image du monde* corresponded to the period identified by scholars of political theory as that of the formation of the nation-state, and in the case of France in particular 'a sense of "national" unity based on the unique holiness of the French people and the French king's special role as the defender of Christianity', to quote Chris Jones.[36] This would suggest in turn that for Gossouin it was not so much a case of 'Europe = Christendom', as Connochie-Bourgne suggests, but perhaps more of 'Europe-as-Christendom = France'.

A number of reservations, however, have to accompany this proposal, and have prompted Jostkleigrewe's follow-up investigation. Before discussing his theory, it is worth mentioning some of the reasons for considering alternatives to Connochie-Bourgne's argument. The first is that this reading rests partly on the notion of the medieval interchangeability of the terms 'Saracen' and 'African' in the Middle Ages, as synonyms for 'pagan' and is founded at least partly on the analytical framework of Orientalism, seeing the text in light of a dichotomy between 'self' and 'other' which presents an opposition between Christians and Muslims, wherein Europe is defined by exclusion and opposition.[37] However, Kathy Cawsey has recently argued very

33 For a discussion, see, for instance, Chantal Connochie-Bourgne, 'Pourquoi et comment réécrire une encyclopédie? Deux rédactions de l'*Image du monde*', in *Encyclopédies médiévales: discours et savoirs*, ed. by B. Baillaud, J. de Gramont and D. Hüe, Cahiers Diderot, 10 (Rennes: Presses Universitaires de Rennes et Association Diderot, 1999), pp. 143–54 (pp. 146–47).

34 For instance, in its extensive discussion of eclipses (absent from the Latin), it mentions that which accompanied the Crucifixion; *Image du monde*, ed. Connochie-Bourgne, pp. 883–90, ll. 4586–4818; the description of the biblical eclipse at pp. 887–90, ll. 4719–4818.

35 See Chapter 3.

36 Jones, 'Understanding Political Conceptions in the Later Middle Ages: The French Imperial Candidatures and the Idea of the Nation-State', *Viator*, 42 (2011), 83–114 (p. 86), citing Joseph R. Strayer, 'France: The Holy Land, the Chosen People, and the Most Christian King', in *Medieval Statecraft and the Perspectives of History: Essays by Joseph R. Strayer* (Princeton, NJ: Princeton University Press, 1971), pp. 300–14.

37 Connochie-Bourgne, 'Limites et diversités', p. 56. Megan Cassidy-Welch, 'Space and Place in Medieval Contexts', *Parergon*, 27 (2010), 1–12 (p. 3); cf. Peter Linehan and Janet L. Nelson, 'Introduction' to Linehan and Nelson, ed., *The Medieval World* (London and New York: Routledge, 2001), pp. 7, 12, who conclude that medieval Europe 'had become Europe by negation: by asserting

forcefully that the divisions and dichotomies often perceived in medieval texts might be due partly to the nature of the theoretical frameworks (such as Orientalism) being applied: 'Formulations which focus on race and geography as the primary maker of difference between peoples and cultures in the medieval period [...] consider the dichotomy between East and West as timeless and essential, rather than as situated and contingent'.[38]

Jostkleigrewe's proposed alternative to Connochie-Bourgne's interpretation of *Image du monde*, argues that to see Gossouin's text as merely shifting boundaries of Europe to match cultural or religious territories would not explain why some regions are in his text not merely reassigned, but apparently attributed to multiple parts of the world simultaneously.[39] A purely geographical or even religious reading is thus unsatisfactory, and to understand Gossouin's world map, what Jostkleigrewe terms a 'historico-geographical model' is needed.[40] This model associates the three parts of the

what it was not, and by proscribing assorted Others' (p. 12). Compare also Heikki Mikkeli: 'During the Middle Ages the view emerged of a world divided in principle into Christians and others'; Mikkeli, *Europe as an Idea and an Identity* (New York, 1998), p. 22. For a discussion of the interchangeable use of terms refering to non-Christian peoples in the medieval Insular context, and further references, see Nataia I. Petrovskaia, 'Which "Pagans"?: The Influence of the Crusades on Battle Narratives in Britain, Ireland and Scandinavia', in *Writing Battles: New Perspectives on Warfare and Memory in Medieval Europe*, ed. by Máire Ní Mhaonaigh, Rory Naismith, and Elizabeth Ashman Rowe (London: Bloomsbury Academic, 2020), pp. 147–64.

38 Cawsey, 'Disorienting Orientalism: Finding Saracens in Strange Places in Late Medieval English Manuscripts', *Exemplaria*, 21 (2009), 380–97 (p. 389). For the original theory, see Edward Said, *Orientalism* (New York: Pantheon, 1978). One wonders to what extent the notion of 'Other' and definition by exclusion is a construct dictated by the modern experience (one thinks in particular of the border controls in the Schengen area, which tend to divide travellers into 'EU' and 'Other', although new additional categories for visa-free travellers, residents, and those possessed of 'fast pass' identification increasingly erode the strict boundary of this dichotomy). Compare also Nora Berend's comments on these seeming oppositions: 'The Concept of Christendom: A Rhetoric of Integration of Disintegration?', in *Hybride Kulturen im mittelalterlichen Europa*, ed. by Michael Borgolte and Bernd Schneidmüller (Berlin: De Gruyter, 2010), pp. 51–62 (p. 60). For a reference to the continuous crossing of interreligious boundaries in medieval Europe see Giles Constable, 'Frontiers in the Middle Ages', in *Frontiers in the Middle Ages: Proceedings of the Third European Congress of Medieval Studies (Jyväskylä, 10–14 June 2003)*, ed. by O. Merisalo and P. Pahta, Textes et Etudes du Moyen Age, 35 (Louvain-la-neuve: Féderation Internationale des Instituts d'Études Médiévales, 2006), pp. 3–28 (pp. 17–18). One is also reminded of Eric Hobsbawm's point that while '[t]he original concept of Europe rested on a double confrontation' of Greeks and Persians on the one hand and Greeks and Scythians on the other, and while this is interpreted, with hindsight, 'in the light of subsequent history, as a process of confrontation and differentiation [...] it would be quite as easy to read into it symbiosis and syncretism'; Hobsbawm, 'The Curious History', p. 219.

39 Jostkleigrewe, 'L'espace', pp. 375–76.

40 Jostkleigrewe, 'L'espace', pp. 377.

world (which he terms 'continents') with three distinct historical moments: Old Testament, New Testament, and contemporary.[41] In this reading, since Africa encompasses the world of the New Testament, and thus both the birth of Christ and the Roman Empire, the city of Rome is associated with it.[42] Similarly, multiple attributions of Greece can be then perceived as due to its historical associations with Alexander the Great and the Apostle Paul on the one hand, and the contemporary Byzantine Empire on the other.[43]

The three points of association Jostkleigrewe proposes for Greece would also work well in the 'Three Orients' model, which I proposed some years ago for the analysis of medieval Welsh literature, distinguishing the representations of the Orient in medieval European literature as carrying biblical, classical, or contemporary associations.[44] Based on the *translatio studii et imperii* concept, wherein the biblical and classical Orients are positioned in the past and privileged as location of imperial power, scholarly knowledge, and religious significance, while the contemporary Orient is perceived as a location left behind by the movement of *studii et imperii* (learning and empire) westward. The fact that multiple associations within *Image du monde* can be mapped onto this pattern also, indicates that rather than being a unique historico-geographical model, as Jostkleigrewe interprets it, Gossouin's text rather presents a unique articulation of a model that was more widely spread across medieval Europe.[45] Jostkleigrewe's interpretation of the medieval perception of space as an overlapping rather than an oppositional structure in essense corresponds to the notion proposed in this book, but presented here in mathematical terms and as part of a broader pattern. The geographical areas of the three parts of the world, Europe, Asia, and Africa, as they are described in the *Imago mundi* tradition, should be conceived of as fuzzy sets.

To illustrate the use of this overlapping model we can turn to consider another passage in the *Image du monde*, wherein the use of 'us' and 'them' terminology has been previously interpreted by Connochie-Bourgne as an

41 Jostkleigrewe, 'L'espace', p. 377.

42 Jostkleigrewe, 'L'espace', p. 377.

43 Jostkleigrewe, 'L'espace', p. 377.

44 Petrovskaia, *Medieval Welsh Perceptions of the Orient*, pp. 5–6. The two models complement each other.

45 The multiple associations of a place with different regions echo what Michael Heffernan describes as the 'spatial fluidity of medieval geopolitics'; Heffernan, 'The Changing Political Map: Geography, Geopolitics, and the Idea of Europe since 1500', in *An Historical Geography of Europe*, ed. by R. A. Butlin and R. A. Dodgshon (Oxford: Clarendon Press, 1998), pp. 140–80 (p. 144).

illustration of a sharply delineated binary system, but which can alternatively be read in terms of fuzziness. The passage runs as follows:

Aussi comme il nous samble ci
Qu'il a grans merveilles enqui,
Aussi resamblent ceus de la
Diverssez les chosez de ca
Et mout merveilleusez les font,
Pour ce que veües nes ont.[46]

Just as it seems here to us
That there are great marvels there,
So seem to them there
Different the things of here
And very marvelous they make them
Because they have not seen them.

Rather than necessarily creating a 'self'/'other' dichotomy, this passage suggests an elision of difference between 'us' and 'them'. Its message is that regardless of their location, people in different parts of the world see the marvellous in distant places where they have not been. 'Here' and 'there' are in that respect similar, and 'we' and 'they' in this passage engage in the same type of imaginings.

The reference to the marvels of Europe in this passage opens up the possibility of reading Europe itself as 'other' in this text. The issue cannot be easily dismissed as incidental, for as Connochie-Bourgne observes in her study of the representation of the Orient in the *Image du monde*, the word *diversitez* appears to be particularly dear to that author.[47] *Etrange* '(strange'), *autre* ('other'), and *divers* ('different'), are according to Connochie-Bourgne preferable in that text to terms 'supernatural' or 'extraordinary'.[48] Yet it must be noted that *diversité*, which signifies 'variety', 'diversity', or 'heterogeneity', does not necessarily imply 'Other' in the sense of 'different from the self'.[49]

46 Gossouin de Metz, *Image du monde*, ll. 3113–18; quoted here from Connochie-Bourgne, 'Limites et diversités', p. 57 (the translation here is mine). For the full text, see Gossouin de Metz, *L'Image du monde*, ed. Connochie-Bourgne.

47 Chantal Connochie-Bourgne, 'L'Orient, réalité et discours, dans l'Image du monde', in *Senefiance 11: Images et signes de l'Orient dans l'Occident médiéval* (Aix-en-Provence: Presses universitaires de Provence, 1982), § 18.

48 Connochie-Bourgne, 'L'Orient', § 18.

49 *DMF*, s.v. *diversité*.

This is amply demonstrated in Connochie-Bourgne's analysis of the use of the term in relation to Europe: 'Des diversitez qui sont en Europe et en Aufrique' ('The marvels which are in Europe and in Africa') further in the text.[50] As she points out, Gossouin's interest in the marvellous lies more in Asia, to which he dedicates considerably more space than to Europe and Africa combined (counting the combined section on the 'diversities' of these areas under discussion here).[51] In Caxton's text, this precedes a discussion of the marvels of Europe, introduced with the words 'We haue in thise parties many thinges that they of Asye and of Affryke haue none'.[52] This links the tradition of the marvels of the East to that of the marvels of the West. The fact that some marvels are present in Europe, in both the French text and in its Middle English translation, makes it difficult to sustain the idea of contrast and 'othering' of Asia through association with marvels. Europe, in these encyclopedias, is as weird and alien. In terms of fuzzy set theory, we can articulate this as follows: we can consider mavels to be a set. However, this set belongs both (partially) to the set that is Asia and partly to the set that is Europe. Thus, marvels form the intersection, in Zadeh's definition of intersection already quoted, between the fuzzy sets Europe and Asia.[53]

The 'marvels of the West' tradition is commonly associated with Britain and Ireland, and it is worth examining the final chapter of the Europe section

50 Connochie-Bourgne, 'L'Orient', § 21. The 'Self–Other' dichotomy derived mainly from the theoretical writings of Jacques Lacan and Emmanuel Lévinas. Examples of the use of this dichotomy contemporary readings of medieval literature and history are too numerous for a full bibliography to be practicable here, but for recent examples of studies building on the theoretical work of Lacan and Lévinas, see, for instance, Michael Uebel, ed., *Ecstatic Transformation: On The Uses of Alterity in the Middle Ages*, The New Middle Ages (New York, NY: Palgrave Macmillan, 2005); Albrecht Classen, ed., *Meeting the Foreign in the Middle Ages* (New York: Routledge, 2002); Siobhain Bly Calkin, *Saracens and the Making of English Identity: The Auchinleck Manuscript*, Studies in Medieval History and Culture (New York and Abingdon: Routledge, 2005); essays in J. C. Frakes, ed., *Contextualizing the Muslim Other in Medieval Christian Discourse*, The New Middle Ages (New York: Palgrave Macmillan, 2011). The major sources for this theoretical approach are Emmanuel Lévinas, *Totalité et infini: Essai sur l'extériorité* (The Hague: Nijhoff, 1971); Jacques Lacan's writings, including, for instance, *Livre XVI: D'un Autre à l'autre, 1968–1969*, ed. by Jacques-Alain Miller (Paris: Éditions du Seuil, 2006). For more on Lacan and a further bibliography of relevant theoretical writings, see Lorenzo Chiesa, *Subjectivity and Otherness: A Philosophical Reading of Lacan* (Cambridge, MA: MIT Press, 2007); and Jean-Michel Rabaté, ed., *The Cambridge Companion to Lacan* (Cambridge: Cambridge University Press 2003).

51 Connochie-Bourgne counts 778 lines for Asia to 252 for Europe and Africa combined; Connochie-Bourgne, 'Limites et diversités', p. 50.

52 Caxton, *Mirrour of the World*, ed. Prior, p. 50.

53 See in this book, p. 31. '[T]he intersection of *A* and *B* is the *largest* fuzzy set which is contained in both *A* and *B*'; Lotfi A. Zadeh, 'Fuzzy Sets', *Information and Control*, 8 (1965), 338–53 (p. 339).

in the Latin text, Chapter 29 *Britannia*.[54] For Britain, no biblical or classical references are possible (to echo Jostkleigrewe's historico-geographical model and the 'Three Orients' model) because it does not feature in the Bible or, for instance, the Alexander legend. This echoes the problem of the nomenclature for designating regions in *Imago mundi*, where *provincie* could only be those areas which had been part of the Roman Empire.[55] While in terms of nomenclature, Britain does not pose the same problem, its position in the farthest west means that in the historical view of *translatio studii et imperii* it is too new to contain any 'history' which might provide a narrative for the geographical locations (in etymologies, for instance).[56] There is awareness in the writings produced in medieval Britain of the marginality of the writers' own location in the medieval worldview: 'Finales hos terrarum angulos' ('these farthest reaches of the Earth'), as Gerald of Wales describes the region at the end of the twelfth century.[57]

54 Flint, ed., *Imago mundi*, pp. 62–63. For discussions of the 'marvels of the West' tradition, see, for instance, Asa Simon Mittman, 'The Other Close at Hand: Gerald of Wales and the "Marvels of the West"', in *The Monstrous Middle Ages*, ed. by Bettina Bildhauer and Robert Mills (Toronto and Buffalo: University of Toronto Press, 2003), pp. 97–112; Yu Onuma, 'Convention through Innovation: Marvels in *Topographia Hibernica* by Gerald of Wales', in *Aspetti del meraviglioso nelle letterature medievali: Aspects du merveilleux dans les littératures médiévales: Medioevo latino, romanzo, germanico e celtico*, ed. by Franca Ela Consolino, Francesco Marzella and Lucilla Spetia (Turnhout: Brepols, 2016), pp. 69–80; Keith Busby, *French in Medieval Ireland, Ireland in Medieval French: The Paradox of Two Worlds* (Turnhout: Brepols, 2017), pp. 191–264.

55 See Chapter 3.

56 That more historical information is given for Asia than for Europe or Africa in the *Image du monde* is observed by Jostkleigrewe, 'L'espace', p. 372.

57 Gerald of Wales, *Descriptio Kambriae*, in *Giraldi Cambrensis Itinerarium Kambriae et Descriptio Kambriae*, ed. by J. F. Dimock, Giraldi Cambrensis Opera, 6 (London, 1868), Praefatio Prima, p. 168 (my translation); note that the reference to remoteness is omitted in Thorpe's translation of the relevant passage: Gerald of Wales, *Descriptio Kambriae*, trans. by Lewis Thorpe, *The Journey Through Wales and the description of Wales* (Harmondsworth: Penguin, 2001), p. 212. The topos has been much commented on; see, for example, Mittman, 'The Other Close at Hand', p. 97; Kathy Lavezzo, *Angels on the Edge of the World: Geography, Literature, and English Community, 1000–1534* (Ithaca, NY: Cornell University Press, 2006), for instance, pp. 2, 7, 54, 85; Roger Ellis, 'Translation and Frontiers in Late Medieval England: Caxton, Kempe, and Mandeville', in *Frontiers in the Middle Ages*, ed. by Merisalo and Pahta, pp. 559–83 (p. 582); Lucy E. G. Donkin, '"*Usque ad ultimum terrae*": Mapping the Ends of the Earth in Two Medieval Floor Mosaics', in *Cartography in Antiquity and the Middle Ages: Fresh Perspectives, New Methods*, ed. by Richard J. A. Talbert and Richard W. Unger, Terminology and Change in History, 10 (Leiden and Boston: Brill, 2008), pp. 189–217, esp. p. 198; Eckhardt, 'One Third of the Earth?', pp. 313–14; and for a more general discussion, see also Anna-Dorothee von den Brincken, *Fines Terrae: Die Enden der Erde und der vierte Kontinent auf mittelalterlichen Weltkarten* (Hannover: Hahnsche Buchhandlung, 1992); and Robert Bartlett, 'Heartland and Border: The Mental and Physical Geography of Medieval Europe', in *Power and Identity in*

Indeed, the application of the Orientalist framework to the analysis of medieval writings about the north-western edges of Europe, the 'marvels of the West', has shown that it is possible to read medieval descriptions of areas and peoples within 'Europe' (both in its modern sense and in the definition of the *Imago mundi*–type medieval geography) as 'othering'.[58] Thus, 'otherness' is not unique to that which is outside Europe in the texts described in this chapter.[59]

The intepretative difficulties concerning the description of the world in *Image du monde* and Gossuin's changes to information derived from the *Imago mundi* may therefore be due only partly to the uniqueness of the thirteenth-century French intellectual context of production, or the specifics of Gossouin's own perceptions of Europe and its place in the world. They may also be more a product of our own changing ideas about the concept of boundary, the function of the Mediterranean as an economic and cultural system, and the inheritance of twentieth-century theories of cultural dichotomies. The changes Gossouin makes seem all the more confusing because they disentangle his tripartite scheme from the Mediterranean and thus from the convenient 'lines' forming the T on the T-O maps. Yet the problem might be less to do with Gossouin's seeming subversion of the T-O structure and more to do with our expectations, formed by the normativity of linear, rather than fuzzy, logic. It would be inadvisable to argue that the 'lines' of the T distinguish political or cultural entities and can be translated into our reading of the textual monuments of medieval geography as borders. Rather, they represent a structural view of the world where the notional components were not necessarily stable or permanent and certainly did not have to correspond to or indicate political boundaries. It may be that Caxton shares our problem because his work post-dates the

the Middle Ages: Essays in Memory of Rees Davies, ed. by Huw Pryce and John Watts (Oxford: Oxford University Press, 2007), pp. 23–36. The perception of Britain as marginal or as part of a different world is already present in Roman literature; see James S. Romm, *The Edges of the Earth in Ancient Thought: Geography, Exploration, and Fiction* (Princeton, NJ: Princeton University Press, 1992), pp. 140–41.

58 See, for instance, Mittman, 'The Other Close at Hand'; Matthieu Boyd, 'Celts Seen as Muslims and Muslims Seen by Celts in Medieval Literature', in *Contextualising the Muslim Other in Medieval Christian Discourse*, ed. by Jerold C. Frakes (New York: Palgrave Macmillan, 2011), pp. 21–38; Onuma, 'Convention', pp. 69–80.

59 Since both *Imago mundi* and *Image du monde* were the progenitors of multiple further texts, it is worth considering this as a starting point for querying other texts as well. The task of investigating the perception of the 'other' in the *Imago mundi* tradition is a monumental enterprise, and the suggestion tentatively offered here can be considered at most a prolegomenon to such an investigation.

development of the linear frontier in the thirteenth century.[60] If so, the notion that the geography of the *Imago mundi* and *Image du monde* is fuzzy, and our evidence, works well with what is known about the development of the idea of the linear border.

The conceptual framework of the historico-geographical model proposed by Jostkleigrewe permits overlap in the various regional schemes and works well with the fuzzy set model proposed in this book. We can test this further by analysing the variation in the description of islands in the *Imago mundi* texts. Viewed through the prism of Jostkleigrewe's model, the fact that the islands of the Mediterranean are often treated separately may be interpreted as a tribute to the plurality of possible attributions of these islands.[61] Due to their position in the T of the T-O world, they can be ascribed to any one of the three regions, and sometimes might be seen to belong simultaneously to several. The next section of this chapter, therefore, looks at what happens to this dataset when it is examined in terms of fuzzy sets.

The Mediterranean as Fuzzy Set

It is something of a commonplace in scholarship on medieval geography and the development of the idea of Europe that the Don boundary was problematic.[62] The evidence of the *Image du monde* suggests that the Mediterranean was, as well. Fuzzy sets help us understand how this might work without being paradoxical. The concept of each part of the world as a fuzzy set allows for areas of overlap, where we are dealing with unities which are members of both sets (perhaps to different degrees). Let us begin by looking at the position of Spain on the margins of both Europe and the Mediterranean, and then we will conclude by examining the Mediterranean itself as a fuzzy set.

We must start this discussion by going back to the list of of Spanish regions in the *Imago mundi* and *Semeiança del mundo* discussed in Chapter 3. The identification of the last region on this list, *Tinguitania*, is uncertain and problematic in a very telling way: it may refer to the Roman province of Mauretania Tingitana, corresponding roughly to modern Morocco.[63] If

60 'Preface' to *Medieval Frontiers*, ed. by Abulafia and Berend, p. xiii; see also above, pp. 29–30.

61 Latini's *Trésor* is a notable exception, but it is loosely and partially related to the *Imago mundi* and the *Image du monde*.

62 See references on p. 96, n. 48 above.

63 Mauretania Tingitania also occurs in Isidore's *Etymologies*, although it should be noted that the list of Spanish provinces in the *Imago mundi*, and in the *Semeiança*, does not quite correspond

so, both the Latin and the Spanish texts incorporate part of what we now consider the African 'continent' into their concept of Europe, emphasising that the boundaries between the three parts of the world did not have to correspond to bodies of water. If the identification is correct, it forms a precedent for the fuzzy map of the *Image du monde*. Both of these examples point to a greater complexity in the medieval tripartite world structure than is usually represented in modern discussions in their use of the formula: three parts of the world separated by the Mediterranean, the Don, and the Nile. They imply a fuzziness and overlap of the sets involved.

The marginality of Spain as member of the Europe fuzzy set in these texts reflects the cultural and political tensions of medieval Spain's 'European' status. As Klaus Oschema points out, medieval authors in general placed Spain unequivocally within Europe.[64] However, this positioning was not devoid of ambiguity. One cannot do better here than cite Sharon Kinoshita's illuminating articulation: 'The dictum that "Europe ends at the Pyrenees" becomes a self-fulfilling prophecy: the Iberian peninsula is not "European" because "Europe" is defined as a culture much smaller than the continent bearing this name'.[65] There are two possible interpretations of the resulting apparent paradox of Spain being both European and not. One of these is to say that the paradox emerges primarily through discursive conflation of 'Europe' as geographical area (our focus here) and Europe as myth and cultural construct (or produced space, to use Lefebvre's terminology).[66] The alternative position, and one that I propose as an extension of the general argument in this book, is that the paradox is only there if one insists on linear boundaries. Within the framework provided by the medieval texts themselves, however, regions and boundaries between them are fuzzy. Spain thus effortlessly shifts in its marginal position as member of the fuzzy set of Europe.

As a result, Spain, locationally within Europe but conceptually problematic and ambiguous, becomes particularly useful as a point of departure

to Isidore's; for a discussion, see from Leszek Wojciechowski, 'Geographical Descriptions of Spain in Latin Medieval Encyclopedias: From Isidore of Seville to Vincent of Beauvais (VII-XIII C.)', *Roczniki Humanistyczne*, 67 (2019), 31–55 (pp. 44, 45).

64 Klaus Oschema, *Bilder von Europa im Mittelalter*, Mittelalter-Forschungen, 43 (Ostfildern: Thornbecke, 2013), pp. 257–60.

65 Sharon Kinoshita, 'Discrepant Medievalisms: Deprovincializing the Middle Ages', in *Worldlings: World Literature, Field Imaginaries, Future Practices: Doing Cultural Studies in the Era of Globalization*, ed. by R. Wilson and Chris Connery (Santa Cruz: New Pacific Press, 2006), pp. 75–89 (p. 86).

66 See above, pp. 19, 30 n. 59.

for examining how the concept of Europe might transform in translation to its margins.[67] In Sharon Kinoshita's words, 'medieval Iberia is in fact a privileged site from which to disrupt reductive notions of the "European" Middle Ages'.[68] The idea is echoed by Julian Weiss, who describes Spain as the 'stereotypical chronotopic other'.[69] It is also interesting that a particular fluidity has been observed in the tripartite structure of the world in Spanish texts such as the *General Estoria*.[70] The uncertainty surrounding the division of Africa and Europe, sometimes separating Africa and Ethiopia, sometimes merging Europe and Africa, may be due to Spain's own cultural and political history.[71]

The different degrees of membership allowed for fuzzy sets also help conceptualise the removal of Britain to the section on Islands in this text and in some of the other *Imago mundi* adaptations. The movement of Spain to the edge of Europe in the *Semeiança* translation of I.21 is an example.[72] In addition to the addition of a reference to Spain, the Castilian translator(s) introduce recapitulating transitional passages between sections of the text.[73] Whereas the final chapter of the Europe section in the *Imago mundi*

67 According to Oschema, no references to Europe are found in ninth- and tenth-century Spanish texts, with Spain instead positioned between Africa and Gaul; Oschema, *Bilder von Europa*, p. 151. Note, however, that – as Oschema observes – even in medieval works that treat Roman provinces with uncertainty, the European position of Spain is not questioned (pp. 260–1). The only exception is the *Image du monde* (n. 116).

68 Kinoshita, 'Discrepant Medievalisms', p. 85.

69 Weiss, 'Remembering Spain in the Medieval European Epic: A Prospect', in *Locating the Middle Ages: The Spaces and Places of Medieval Culture*, ed. by Julian Weiss and Sarah Salih (London: King's College London Medieval Studies, 2012), pp. 67–82 (p. 75).

70 Aníbal A. Biglieri, *Las ideas geográficas y la imagen del mundo en la literatura española medieval*, Medievalia Hispanica, 17 (Madrid: Iberoamericana; Frankfurt am Main: Vervuert, 2012), pp. 62–63.

71 Biglieri, *Las ideas geográphicas*, pp. 62–64.

72 William E. Bull and Harry F. Williams, eds, *Semeiança del mundo: A Medieval Description of the World* (Berkeley and Los Angeles: University of California Press, 1959), p. 78. Compare the *Primera Crónica General de España*, 2 vols., ed. by Ramón Menéndez Pidal (Madrid: Editorial Gredos–Cátedra Seminrio Menéndez Pidal, 1979), I, 5–51; quoted and discussed in Biglieri, *Las ideas geográficas*, pp. 65–66. An important element common to texts of the *Imago mundi* tradition, retained here, is the alternative etymology referring to Europs; Bull and Williams, eds, *Semeiança del mundo*, p. 16; Alan Deyermond, 'Building a World: Geography and Cosmology in Castilian Literature of the Early Thirteenth Century', *Canadian Review of Comparative Literature*, 23 (1996), 141–59 (pp. 143–44); and references above, p. 95, n. 47.

73 Bull and Williams, eds, *Semeiança del mundo*, p. 16. It is worth noting that Spain is mentioned by Isidore in his description of the extent of Europe. Whether the author is following Isidore here, or simply introducing a reference to Spain as subject of particular interest to his audience, is a moot point. For instance, a passage recapitulating the information of a previous section is

is Britain, in *Semeiança* it is Spain (Chapter 128). Britain is moved to the later, separate discussion of islands. This act breaks the original design of the *Imago mundi*, where the separate section was dedicated to the islands of the Mediterranean. The removal of Britain from the concluding position in the description of Europe in the *Semeiança* puts Spain in a marginal position on the European map of this text, echoing the structure of Isidore's description of Europe.[74] It also suggests that either for the Spanish translators the section on Islands in *Imago mundi* did not seem to be a geographically coherent unit (the British Isles and Ireland are, after all, not in the Mediterranean) or that they were thinking in *mappae mundi* terms, where these islands are not that far removed from the Mediterranean.

Looked at through Zadeh's mathematical model, the Mediterranean itself constitutes a fuzzy set.[75] The category of 'Islands' in the *Imago mundi* follows the chapters dedicated to the three parts of the world (1.33–35). This category thus does not fall into the sets of Asia, Africa, or Europe, and the discussion in the section of the text focuses primarily on islands of the Mediterranean (the British Isles and Ireland having been included in the part of the text devoted to Europe).[76] The Islands covered in Chapter 33 are: Cyprus, Crete, Abydos,[77] Cos, Cyclades, Rhodes, Tenedos, Karpathos, Cythera, Delos, Ikaria, Naxos, Melos, Paros, *Cidon*,[78] Samos.[79] Chapter 34 is dedicated to Sicily, with reference made to the Aeolian Islands and the Stoechades.[80] Chapter 35

added to the end of the section on Europe (as Chapter 129): *Aqui se acaba tierra de Europa* ('Here ends the land of Europe'); Bull and Williams, eds, *Semeiança del mundo*, p. 88.

74 For more on Isidore's description of Spain, see Wojciechowski, 'Geographical Descriptions', pp. 37–40.

75 See discussion above, pp. 121–22.

76 The treatment of islands as a separate category is a feature of other texts in the *Imago mundi* tradition, including the *Image du monde*, and follows Isidore; see the discussion in Bernard Ribémont, 'L'inconnu géographique des encyclopédies médiévales', *Cahiers de recherches médiévales*, 3 (1997), 101–11 (pp. 107–108). Brunetto Latini diverges from this trend in incorporating islands into the T-O structure; see ibid., p. 108. For further discussion of the structure of Latini's geographical descriptions, see below, pp. 151–53. For discussion of Britain as a boundary case, sometimes depicted as part of Europe and sometimes as external, see also Wintle, *The Image of Europe*, p. 36.

77 City on the mainland, mistakenly identified as an island in the text.

78 Identification unclear, but associated in the text with the production of *mastix*, mastic gum, a natual resin traditionally produced on the Greek island of Chios; see Nikolas Bakirtzis and Xenophon Monairos, 'Mastic Production in Medieval Chios: Economic Flows and Transitions in an Insular Setting', *Al-Masāq: Journal of the Medieval Mediterranean*, 31 (2019), 171–95.

79 Flint, ed., *Imago mundi*, pp. 64–65.

80 Flint, ed., *Imago mundi*, p. 65. Based on the variation in the references to Sicily and Samos in the corresponding section of the *Image du monde*, Christine Silvi has found that for that part of the text at least, the prose version of French text is closer to the original Latin, suggestive

mentions Sardinia, Corsica, Cyrnus, Ebosus, *Colubria* (= Formentera?), the Balearic Islands, *Gorgones insulae* (=Cape Verde?), the Hesperides, Atlantis (unnamed but described), Island of Meroë, and finally *Perdita*, the name of which translates as 'lost', and which is associated with the legend of the seafaring Irish saint, St Brendan.[81] It will be observed that, with the possible exception of the 'lost' and therefore by definition unlocated *Perdita*, most of the islands discussed in the separate section in the *Imago mundi* are in the Mediterranean.[82] This may indicate that the discussion of landmasses in the Mediterranean under a separate heading, as 'Islands' points to the perception of the sea as fuzzy set 'the Mediterranean'. This is also indicated by the transitional phrase at the conclusion of the chapter on Africa (Chapter 32), which introduces the discussion of the islands (*peragratis Africę finibus, ad insulas maris tendamus*; 'having travelled the ends of Africa, we would press on to the islands of the sea'), is similar to the transitional phrases found at points when the discussion moves from Asia to Europe and from Europe to Africa.[83]

In this regard it is important to remind ourselves that the British Isles and Ireland, though technically also islands, are placed in the Europe section of the text (they belong to the fuzzy set Europe). The only potential problem for this interpretation is that posed by the use of word *dirimitur* ('divide') in relation to the Mediterranean. It would, at first glance, seem to imply that the use of Mediterranean (islands) as a separate fuzzy set (and a separate section in the text) does not correspond to the structure Honorius sets out at the beginning of Book I, which specifies that the inhabited world is composed of three parts. However, as shown in the beginning of this chapter, the term does not have to imply a linear boundary. It can be used of a region that falls between other regions and separates them. If this interpretation

of a process of correction in reference to the Latin or the use of a source different to the verse version; see Silvi, 'Les variations sur le nom dans la *Mappemonde* de maître Gossuin de Metz', in *Par les mots et les textes. Mélanges de langue, de littérature et d'histoire des sciences médiévales offerts à Claude Thomasset*, ed. by Danielle Jacquart, Danièle James-Raoul, and Olivier Soutet (Paris, 2005), pp. 679–91 (p. 683).

81 For a discussion of the afterlife of the Irish Brendan legend in later – post-medieval – works, and further references, see Natalia I. Petrovskaia, *Modern Afterlives of Old Irish Travel Narratives: From Gulliver to Star Trek* (Leiden: Brill, forthcoming).

82 Perhaps this reference to Perdita (one of the 'Fortunate Isles') in *Imago mundi* is the reason for the Brendan story to be drawn into the *Image du monde*. For more on the 'Fortunate Isles', see, for example, Vincent H. Cassidy, 'Other Fortunate Islands and Some that were Lost', *Terrae Incognitae*, 1 (1969), 35–39; and Marcos Martínez Hernández, 'Las Islas Afortunadas en la Edad media', *Cuadernos del CEMyR*, 14 (2006) 55–78.

83 Flint, ed., *Imago mundi*, p. 64; compare ibid., pp. 59, 63.

is correct, it also solves the seeming inconsistency in Honorius's design. Where in the traditional interpretation it looked as if he set out dividing the world into three, and then proceeding to describe four parts of it, in this interpretation he is setting out four regions in the first place: the three parts of the world (Asia, Europe, Africa), divided by a fourth (the Mediterranean). We may therefore need to consider it as a separate fuzzy set in its capacity as a fourth 'part of the world', which in consequence should be described not as a tripartite, but as a quadripartite one, with considerable overlap.[84]

The Mediterranean provides a particularly apt illustration for why fuzzy sets are a helpful heuristic tool. Studies focusing on the cultural world of the Mediterranean and surrounding areas have shown the need for the re-evaluation of what for a long time had been considered as borders and boundaries.[85] As Eric Hobsbawm observes: it is possible 'to see the entire Mediterranean civilization of classical antiquity as syncretic', and it is that civilisation that gave birth to the tripartite conception of world geography inherited in the Middle Ages.[86] For the classical world, Hobsbawm continues, 'the present division between Europe, Asia and Africa had no meaning – at least no meaning corresponding to the present [...]. What meaning could it have had in the heyday of the undivided Roman Empire, happily tricontinental and ready to assimilate anything useful that came from anywhere?'[87] That there was no break between the connected Mediterranean of the classical period and the connected Mediterranean of

84 For more on the Mediterranean as a unit, and a connecting rather than a dividing force, see the studies cited in n. 85 below; see also discussions in Brian A. Catlos and Sharon Kinoshita, eds, *Can We Talk Mediterranean? Conversations on an Emerging Field in Medieval and Early Modern Studies* (Basingstoke: Palgrave Macmillan, 2017).

85 See, for instance, Allen James Fromherz, *The Near West: Medieval North Africa, Latin Europe and the Mediterranean in the Second Axial Age* (Edinburgh: Edinburgh University Press, 2016), whose argument is that 'medieval Western European and North African history were part of a common Western mediterranean culture' (p. 1). See, however, David Abulafia's review in *Journal of Islamic Studies*, 29 (2018), 110–12; also Abulafia, *The Great Sea: A Human History of the Mediterranean* (Oxford: Oxford University Press, 2011), esp. p. 241, for the loss of unity of the Mediterranean cultural space in the early Middle Ages. For a much earlier reference to the same idea, see Hobsbawm, 'The Curious History', p. 219. The great predecessor of the studies dedicated to the Mediterranean is, of course, Fernand Braudel's *La Méditerranée et le Monde Méditerranéen à l'Epoque de Philippe II* (Paris: A. Colin, 1949), which in its second movement presents the sea as a historical actor in its own right. For comments on the originality of this presentation of the Mediterranean as a historical character, see the discussion by the work's dedicatee and Braudel's doctoral advisor, Lucien Febvre, in 'Un livre qui grandit: *La Médieranée et le monde méditeranéen a l'époque de Philippe II*', *Revue historique*, 203.2 (1950), 216–24 (p. 218).

86 Hobsbawm, 'The Curious History', p. 219.

87 Hobsbawm, 'The Curious History', p. 219.

the later Middle Ages has also been shown by Michael McCormick.[88] The Mediterranean as a fuzzy set overlapping with the other regions (also fuzzy sets) allows us to combine what McCormick described as the connected map with the T-O map as traditionally understood (but sans linear borders) without doing 'epistemic violence' (to borrow Gayatri Chakravorty Spivak's term) to either.[89]

This chapter began with an investigation of whether the language of the texts permits us to draw border and boundary lines. I have argued that more attention must be paid to the descriptive meaning of the French term *deviser* and that the apparent Europe–Asia boundary drawn at the Dardanelles in a translation of Brunetto Latini's text can be reinterpreted as a fuzzy area of overlap between two fuzzy sets. Fuzzy sets also help us understand, alongside Jostkleigrewe's argument, how several regions described in Gossouin's *Image du monde* can belong simultaneously to different parts of the world. Finally, we returned to the argument for reading the Mediterranean and its islands as a fuzzy set, too. The last can be interpreted both in the framework of Jostkleigrewe's historico-geographical model and of my fuzzy set model. The fuzzy set model for the geographical descriptions was thus applied to four distinct cases. The reader may ask whether we have progressed at all beyond the starting point of this book: that Ruggiero on the hippogriff flying out of Europe must have meant something. If the world of the *Imago mundi* texts is a fuzzy one, how does one navigate it? In the following chapter I aim to answer this question by exploring the means by which the audience journeys through the various regions described in these texts. I argue that this is done through reference to the regions themselves, rather than to their boundaries, and the audience is taken on an armchair journey through fuzzy areas, not across linear borders. Are the Pillars of Hercules, the 'sign' left for mariners that mark the 'ends' of Europe for Ruggiero in *Orlando Furioso* really a border, or are they a central point in the fuzzy area of overlap (a frontier zone) between two fuzzy sets? I have aimed to show in this chapter that, in the *Imago mundi* tradition at least, borders are about as present as

88 Michael McCormick, *Origins of the European Economy: Communications and Commerce AD 300–900* (Cambridge: Cambridge University Press, 2001); McCormick's study is mentioned in the context of the notion of an expanding Europe of the later medieval period in Jace Stuckey, 'Introduction', to *The Eastern Mediterranean Frontier of Latin Christendom*, ed. by Stuckey, The Expansion of Latin Europe, 1000–1500, 6 (Farnham and Burlington: Ashgate, 2014), pp. xvii–xxxv (pp. xix–xx).

89 Gayatry Chakravorty Spivak, 'Can the Subaltern Speak?', in *Marxism and the Interpretation of Culture*, ed. by Cary Nelson and Lawrence Grossberg (Urbana and Chicago: University of Illinois Press, 1988), pp. 271–313 (p. 280).

the linear shoreline is for one who walks into the sea: with their feet dry one moment and the next submerged in water; yet at the same time, the water might recede, offering another second's respite, leaving the sand behind them wet. Where does the sea start?

5. *Movement*. The Hodoeporical Descriptive Technique

Abstract: This chapter examines the order in which the various regions are described. The focus is on the re-ordering of the regions of Europe in several adaptations of the *Imago mundi*, which serve to refocus the text on the region in which the adaptation was produced. This is examined in the context of changes to the order of regions of Italy in different versions of the *Imago mundi* itself, demonstrating that the text was consciously designed to walk the reader through the world (almost literally). This yields a particularly flexible itinerary framework for the geographical information conveyed, and thus facilitates such re-ordering in the vernacular translations and adaptations. To describe this phenomenon, the term 'hodoeporical descriptive technique' is introduced in this chapter.

Keywords: aurality, reader experience, itinerary, hodoeporical descriptive technique, Fazio degli Uberti, Brunetto Latini

Having visited the islands of the Mediterranean at the end of Chapter 4, we remain in the vicinity for now, circling on our metaphorical hippogriff over the regions of Spain and Italy. I mentioned at the beginning of our journey in the context of discussion of linear borders that we might see Ruggiero's journey on the hippogriff as not crossing borders but entering regions. This was part of my argument for fuzzy sets. In Chapter 4 we looked at geographical areas described in our tradition as fuzzy sets, as part of our investigation into the why of *Imago mundi*'s popularity. In this chapter I want to move towards investigating how the text was experienced. We will first use the Spanish translation of *Imago mundi* to explore the relationship between the text and the audience's experience thereof. This will touch on issues of aurality and raise some questions (though perhaps fewer answers) about the performative aspects of this encyclopedic material. Then, we will look at how the audience is guided through the geography as an armchair

Petrovskaia, N.I. *Transforming Europe in the Images of the World, 1110–1500. Fuzzy Geographies.* Taylor & Francis Group, 2025
DOI 10.5117/9789048563166_CH05

journey, and how that journey changes in the various vernacular adaptations of our Latin text.

Reader, Listener, or Viewer?

In introducing Europe, the Spanish text also recapitulates preceding information:

> Ia oystes de suso como se departe el mundo en tres partes: en Asia e en Europa e en Affrica. Ya oystes de Asia la maior, que es en tierra de India, e Asia la menor, que es en tierra de Bitinia, e terra de Frigia.[1]

> You have already heard how the world is divided into three parts: into Asia and into Europe and into Africa. You heard already of Asia the Greater, which is in the land of India, and Asia the Lesser, which is in the land of Bitinia, and land of Phrygia.

The association of Asia the Greater with India is reminiscent of the relabelling of Asia as India in *Image du monde*, though such drastic rewriting is not present in the *Semeiança*.[2] The transitional passage connects the two regions (and sections), and reminds the audience of what had gone before, much like a voiceover in a modern TV series might at the beginning of an episode.[3] I have also been doing it in this book. The translator(s) add a new transitional chapter after the Chapter 12 (at this point the *Imago mundi* simply starts the next chapter, 13, with *Ab Indo flumine usque ad Tigrim...*):[4]

1 William E. Bull and Harry F. Williams, eds, *Semeiança del mundo: A Medieval Description of the World* (Berkeley and Los Angeles: University of California Press, 1959), pp. 76, 78. The translation is mine.

2 Above, p. 111 n. 23.

3 Natalia I. Petrovskaia, 'Re-Writing Parts of Europe in Some Vernacular Adaptations of the *Imago Mundi*', in *Medieval Translatio. Interdisciplinary Studies in the Translation and Transfer of Language, Culture, Literature*, ed. by Massimiliano Bampi and Stefanie Gropper (Brelin: DeGruyter, 2024), pp. 81–101 (p. 95).

4 Honorius Augustodunensis, *Imago mundi*, ed. by Valerie I. J. Flint, *Archives d'histoire doctrinale et littéraire du Moyen Âge*, 49 (1982), p. 55. It may be worth pointing out that Gossouin in his *Image du monde* also seemed to have had trouble with this rapid transition, as his text at this point appears to fail to switch to the next area, resulting in the reassignation of the Middle East to India, which can alternatively be analysed as the relabelling of the whole region of Asia as 'India'; see Natalia I. Petrovskaia, 'Mythologizing the Conceptual Landscape: Religion and History in *Imago Mundi, Image du Monde* and *Delw y Byd*', in *Landscape and Myth*, ed. by

> Ia oystes de suso de tierra de India como era rica [...], e oystes otro si del mar que corre por ella e dalgunas sos yslas e de los pescados e de las bestias e de las gentes que moran y fata que allegamos al rrio que ha hy, que dizen Indus.[5]
>
> You have already heard above of the land of India, how it was [...], and you heard apart form that of the sea which runs through it and of some of its islands and of the fish and the animals and the peoples who live there until we came as far as the river which is there, which they call Indus.

These passages appear to guide the audience. I use the term 'audience' because, while the text does seem to refer to the visual aspect of reading (through use of *de suso*, 'above'), *oyestes* ('you heard'; < *oír* 'to hear') indicates oral/aural delivery.[6]

The problem of oral or written transmission or delivery in Spanish texts has been much discussed, and the only consensus at present seems to be caution in drawing any conclusions on the basis of such clearly formulaic expressions.[7] It is worth noting, however, that no such terminology is used in the original Latin *Imago mundi*. There are no indications of aurality in the Latin text. The Castilian text thus appears to change the intended reception mode for the text and the information it contains.

The *Imago mundi* itself rather constructs a virtual journey, where the armchair traveller reads through the world in a sequence of geographical juxtapositions, as we have seen in the passages quoted in the previous chapters.[8] Words such as *transeamus* (pres. sj. 1 pl. < *transire*, 'go over') indicate a mental perambulation or virtual travel.[9] We have already seen

Matthias Egeler, Medieval Texts and Cultures of Northern Europe (Turnhout: Brepols, 2019), pp. 195–211 (pp. 203–06); 'Re-Writing Parts of Europe', pp. 95–96.

5 Bull and Williams, eds, *Semeiança del mundo*, p. 62.

6 Petrovskaia, 'Re-Writing Parts of Europe', p. 96. The difficulty of the apparent confusion between written and oral in such texts has been acknowledged for a long time. For an early discussion, see José Amador de los Rios, *Historia Critica de Literatura Española*, 7 vols (Madrid: José Rodriguez, 1863), III, p. 605. See also Steven N. Dworkin, *A Guide to Old Spanish* (Oxford: Oxford University Press, 2018), p. 15; Roger M. Walker, *Tradition and Technique in 'El Libro del Cavallero Zifar'* (London, 1974), pp. 145–46; G. B. Gybbon-Monypenny, 'The Spanish *Mester de Clerecía* and its Intended Public: Concerning the Validity as Evidence of Passages of Direct Address to the Audience', in *Medieval Miscellany Presented to Eugène Vinaver*, ed. by F. Whitehead, A. H. Diverres, and F. E. Sutcliffe (Manchester: Manchester University Press, 1965), pp. 230–44 and the early study in Ruth Crosby, 'Oral Delivery in the Middle Ages', *Speculum*, 11 (1936), 88–110.

7 See references above.

8 See discussion above, pp. 95, 98, 102.

9 See above, p. 127.

a similar phenomenon in the *Image du monde* of Gossouin de Metz. In the introduction, this text employs the first-person verbal form, but as Simon Gaunt points out 'this "nous" is strictly academic'.[10] The address to the audience is not intended, as in the Spanish text, for oral performance, but is rather a convention referring to the joint armchair journey of the audience or reader.

The shift we therefore observe in the *Semeiança* seems to be reflective of a shift in genre almost, and a process of cultural translation: not only being translated into a new language, and into a new political context, with new information added concerning Spain, but also a new performative format, meeting the requirements and conventions of the new audience (and host culture).[11]

This phenomenon has been observed elsewhere in medieval translations, and in different genres. The translator responsible for the Welsh *Ystorya Bown o Hamtwn* (Story of Bown of Hampton), for instance, 'implemented major changes on the levels of narrative structure and style', to quote the text's editors.[12] This reflects the more creative pattern followed by medieval Welsh translators in narrative texts as compared to, for instance, *Delw y Byd*, the medieval Welsh translation of *Imago mundi*, which is a faithful rendering of the original (including the hodoeporical terminology that we will come back to in the second part of this chapter).

The transitional passages introduced in the Spanish translation might imply that the reader (whether reading for an audience or for themselves) will have made a pause at the end of the previous section and requires a recap, as a modern television viewer might. These suggestions must, however, be treated with caution, as the formulaic exhortation to an audience is typical of Spanish texts of the period.[13]

A search for historical uses of the term *oyestes* (in the 2sg. pret. form) in the Corpus diacrónico del español (CORDE), yields 543 instances in 41 texts, ranging in date from the thirteenth to the fourteenth centuries (with two

10 Simon Gaunt, *Marco Polo's 'Le Devisement du Monde': Narrative Voice, Language and Diversity* (Cambridge: D. S. Brewer, 2013), p. 70. See also discussion in Petrovskaia, 'Re-Writing Parts of Europe', p. 96.

11 Petrovskaia, 'Re-Writing Parts of Europe', pp. 96–97.

12 Erich Poppe and Regine Reck, 'A French Romance in Wales: *Ystorya Bown o Hamtwn*. Processes of Medieval Translations (2)', *Zeitschrift für celtische Philologie*, 56 (2008), 129–64 (pp. 142, 147–49). For the edition, see Poppe and Reck, eds, *Selections from Ystorya Bown o Hamtwn* (Cardiff: University of Wales Press, 2009).

13 See Dennis P. Seniff, 'Orality and Textuality in Medieval Castilian Prose', *Oral Tradition*, 2 (1987), 150–71 (pp. 151–53).

undated instances).[14] By far the predominant genre of text in which the term occurs is historical prose (361 instances, 66.48%), followed by narrative prose (67 instances, 12.33%), religious prose (49 instances, 9.52%), and scientific prose (30 cases, 5.52%). Verse narrative and lyric genres produce even fewer examples. It must be added that all but one of the instances of the term *oystes* in 'scientific prose' are from the *Semeiança*, and only one comes from the *Lapidario*.[15] These statistics should be treated with caution, therefore, as a single long text with frequent formulaic use of the term might affect the numbers considerably. A large proportion of the cases of the term's use in historical prose, for instance, come from the monumental encyclopedic universal history composed in the reign of Alfonso X, the *General Estoria*.[16] Nevertheless, the pattern which shows dominant use in texts which provide a progressive narrative or story (historical prose and narrative texts) is unlikely to be an accident. If so, the *Semeiança* rewrites the Latin text stylistically to fit in that context. Its description of space seems to be subject, in its Castilian reworkings, to the same narrative conventions as other texts' descriptions of time. We are back to space-time and the *chronotope*. This might also lead us to question not only the performative aspect of the text, but more crucially our preconceptions of its genre. The genre distinctions in CORDE are modern. The question therefore is whether the *Semeiança* should, in fact, be grouped with the other texts, and if so, according to which criteria. We could be dealing with a common, broader, historico-geographical genre.

To what extent phrases such as *ia oystes de suso* reflect actual practice remains an important question, but one which lies beyond the scope of the present study. To further elucidate this issue an analysis of the use of terminology for both conveyance and reception of information in these texts

14 Real Academia Española, Banco de datos CORDE, http://corpus.rae.es/cordenet.html (accessed 6 August 2020).

15 For an edition of this text, completed *c.* 1250, see Alfonso X, *Lapidario (según el manuscrito Escurialense H.I.15)*, ed. by Sagrario Rodríguez M. Montalvo (Madrid: Editorial Gredos, 1981); Lloyd Kasten, John Nitti, and Jean Anderson, *Concordances and Texts of the Royal Scriptorium Manuscripts of Alfonso X, el Sabio* (Madison, WI: Hispanic Seminary of Medieval Studies, 1978). For discussions, see, for instance, Ana Domínguez Rodríguez, *Astrología y arte en el 'Lapidario' de Alfonso X el Sabio* (Madrid: Edilán, 1984).

16 For the *General Estoria*, see *Alfonso X, 'El Sabio': General Estoria, VI Partes*, ed. by Pedro Sánchez-Prieto, 10 vols (Madrid: Biblioteca Castro, Fundación José Antonio de Castro, 2009); for an earlier edition, see Alfonso el Sabio, *General estoria*, Part 1, ed. by Antonio G. Solalinde (Madrid: Centro de Estudios Históricos, 1930) and *General estoria*, Part 2, ed. by Antonio G. Solalinde, Lloyd A. Kasten, and Victor R. B. Oelschläger, 2 vols. (Madrid: Consejo Superior de Investigaciones Científicas, 1957–61).

would be in order, such as the use of *dezir* ('to say'), which frequently occurs in proximity to *oír* ('to hear') in these texts, as well as the more ambiguous *contar* ('recount, narrate'). While such an investigation is beyond the scope of this book, on the basis of the available information it is possible to cautiously conclude that, at the very least in the *Semeiança*, *Imago mundi* material was rewritten to match the conventions of the host vernacular culture, as well as to emphasise the place of Spain in Europe.

Another point arising from the brief presentation here is that caution is in order regarding assumptions of the difference in reception of texts we now consider as belonging to different genres. The versatility of geographical information of the *Imago mundi* tradition, often incorporated into narratives such as Alexander romances suggests that geographical accounts would have been performed as much as read (whatever that latter process might have entailed in practice at the time). At the very least, this is another indication of the 'tense relation between orality and literacy and its impacts on in medieval society' to once more quote Liedeke Plate and Els Rose, whose definition of rewriting I have adopted.[17] A tentative suggestion that can be offered at this point, as a contribution to the debate on orality versus literacy in the reception of Spanish texts, is that the distinction between reading and listening might be a purely modern phenomenon, as it is particularly marked for a public used to silent reading.[18] In the absence of silent reading (or in a culture where it is not the dominant mode of text access), whether an individual was reading with or without an audience becomes irrelevant in terms of auditory reception, because that individual could also have read aloud in the course of solitary perusal.

While Spain as a single case study is not necessarily reflective of the medieval European experience of encyclopedic material as a whole, comparison with other examples yields a firmer indication of an aural culture. A similar phenomenon to that discussed here was observed for medieval Scandinavian literature by Sif Ríkharðsdóttir, who points out that in the conclusion to *Karlamagnus saga ok kappa hans*, the Norse Charlemagne saga,

17 Liedeke Plate and Els Rose, 'Rewriting, A Literary Concept for the Study of Cultural Memory: Towards a Transhistorical Approach to Cultural Remembrance', *Neophilologus*, 97 (2013), 611–25 (p. 622).

18 For more on medieval versus modern reading practices, and the history of silent reading, see Paul Saenger, *Space Between Words: The Origins of Silent Reading* (Stanford: Stanford University Press, 1997); Mary Carruthers, *The Book of Memory: A Study of Memory in Medieval Culture*, 2nd edn (Cambridge: Cambridge University Press, 2008), pp. 212–17; and the discussion of individual readers' practices in D. H. Green, *Women Readers in the Middle Ages* (Cambridge: Cambridge University Press, 2007), pp. 1–15.

the combination of references to written and aural transmission seems to point to an aural reception of the written text.[19] By contrast to the Norse phenomenon, which Sif Ríkharðsdóttir describes as 'sustained emphasis on the textuality of the transmission and its perpetuation' in the course of the shift from aurality to textuality, the Spanish tradition appears to sustain the emphasis on the aural element of the text transmitted in written form.[20] These may be regarded as different approaches towards a similar cultural problem.

Going through the Text

While further research is necessary in order to be able to say conclusively for each text of the *Imago mundi* family whether the audience experience was (intended to be) visual or aural, some observations can be made on the basis of the available evidence regarding the way in which the audience is led through the material (whether by eye or ear). This section of the book thus addresses the descriptive technique of the *Imago mundi* texts, which I call hodoeporical. We will look at how the audience is guided through the text as a whole, and then focus on two regional examples within the description of Europe. The focus here is on the order in which areas are recounted, and the principles underpinning it. One of the reasons why this needs to be addressed is that changes to the sequence within a region have sometimes led scholars to assume that a text was not based on the *Imago mundi*. I argue here that these alterations are often facilitated by the structure of the original treatise and constitute a feature of the process of rewriting in its vernacular adaptations. I also highlight the fact that sequence alterations are found not only in adaptations but also within the Latin manuscript tradition.

19 I follow the distinction made by Sif Ríkharðsdóttir in 'The Norse Roland in Context', paper presented at the Crossing Borders in the Insular Middle Ages Symposium, 24–25 September 2019. For more on the transformations in these texts from the aural to the textual in the course of translation, see Sif Ríkharðsdóttir, 'The Norse Roland in Context', in *Charlemagne in the Norse and Celtic Worlds*, ed by Helen Fulton and Sif Ríkharðsdóttir (Cambridge: D. S. Brewer, 2022), pp. 50–65. I am grateful to Sif Ríkharðsdóttir for generously sharing a pre-publication version of her article. See also Gabriele Röder, 'Die chansons de geste der altnordischen *Karlamagnús saga*: Übersetzungen oder Adaptationen', in *The Medieval Translator: Traduire au moyen âge*, ed. by Roger Ellis, René Tixier, and Bernd Weitemeier, The Medieval Translator, 6 (Turnhout: Brepols, 1998), pp. 134–58.

20 Sif Ríkharðsdóttir, 'The Old Norse Roland in Context', p. 62.

Another reason why it is important to pay attention to the sequence of narration is that the vernacular versions frequently place the region with which they themselves are associated last in the description of Europe, thereby emphasising it. In this chapter I use the example of Italy for the purposes of analysis. It provides rich ground for discussion and comparison in this respect, as it is home to several independent rewritings of our text: Brunetto Latini's *Trésor*, Vivaldo Belcalzèr's *mapa del mond* (which forms part of his translation of Bartholomeus Anglicus's *De proprietatibus rerum*), and Fazio degli Uberti's *Dittamondo*. These adaptations have very different approaches to the text, but tend not to be discussed together. It is useful to see how each takes advantage of the fuzziness of the *Imago mundi* to bend the geographical description to new purposes and make it serve a new audience. Plate and Rose's concept of rewriting is particularly useful in understanding this process, for the changes observed in the *Imago mundi* family are indeed 'an act of cultural memory' and thus constitute precisely a case where 'rewriting asks scholars to account for agency and intentionality'.[21]

This is one of the few cases where one has to start at the end and not at the beginning. Let us begin with the final chapter of the Europe section in the *Imago mundi* (I.29). This as we have seen, is dedicated to Britain in the Latin original. A general pattern is that the various vernacular rewritings of the *Imago mundi* often refocus their text to their own region of origin by placing it last in the sequence of European lands visited.[22] For example, the *Image du monde* adds the mention of France after Britain, to close its discussion of Europe. The same change can be seen in both the *Semeiança* and in Juan de Mena's *Laberinto de Fortuna*. These authors achieve the same results differently, however, for they move Britain and Ireland to the section of islands. Spain, which in the *Imago mundi* was in the penultimate chapter on Europe, I.28, is thus moved to the concluding position.[23]

21 Plate and Rose, 'Rewriting', p. 623; and p. 58 in this book.

22 Jacob van Maerlant also moves the section on British Isles and Ireland – which in Honorius's text follow the reference to Spain – towards the end of the section in Islands, which follows his description of Africa; Jacob van Maerlant, *Alexanders geesten*, ed. by Johannes Franck (Groningen: J. B. Wolters, 1882), pp. 271, 275–76. This latter change appears to be Maerlant's own innovation, rather than his Latin source; Petra Berendrecht, *Proeven van bekwaamheid: Jacob van Maerlant en de omgang met zijn Latijnse bronnen* (Amsterdam: Prometheus, 1996), p. 40. In this case, however, the change Maerlant makes does not result in his own area being last. However, as we have seen in the previous discussion, Maerlant's emphasis on the local region is expressed differently, simply because it wasn't in the *Imago mundi* in the first place – he adds it to the map.

23 Flint, ed., *Imago mundi*, p. 62.

The same trend, to put the host region at the conclusion, is also found outside the *Imago mundi* tradition. The Old Norse Reynistaðarbók, MS AM 764 4to (s. XIV), for example, contains a description of the world, in the sequence Asia–Africa–Europe, and places Iceland at the end of its discussion of Europe (and thus at the end of its discussion of the world as a whole).[24] The text forms part of a chronicle and is not related to the *Imago mundi* tradition, but shows a similar attitude to the rewriting of geographic material.[25] It seems, therefore, that placing the home country last in the list, representing it as the edge of Europe, was a more general convention.

Three possible, and not necessarily mutually exclusive, explanations can be offered for this phenomenon. In the first place, particularly in the case of the chronicle, the home region might have been mentioned last in the text in order to connect the geographical description to the following historical account. This explanation seems to fit with the phenomenon referred to in Chapter 1 of the difference in the order in lists of the three regions in geographical and historical texts.[26] For the historical texts, the order Asia→Africa→Europe seems to follow the order of *translatio imperii*.[27] In the case of of *Imago mundi*–derived texts, however, this is a little more complex, since there Europe was followed by Africa. However, since extracts from these texts were sometimes used in chronicles and other sources, it is not inconceivable that this may have been part of the design also. The second possibility is that the final place was a place of honour and emphasis (this is a reading somewhat edging towards the Eurocentrism concept), echoing the ascending order in religious processions.[28] The third possibility,

24 See Petrovskaia, 'La formulation tripartite', p. 203. For a more in-depth discussion of the geographical description in this text, and further references, see Svanhildur Óskarsdóttir, 'Universal History in Fourteenth-Century Iceland: Studies in AM 764 4to' (unpublished PhD thesis, University of London, 2000), pp. 69–73. For the manuscript, see https://handrit.is/en/manuscript/view/da/AM04-0764 (accessed 20 September 2021).

25 The order in which it describes the world follows a pattern ultimately traceable to Orosius rather than to Isidore (as in the *Imago mundi* texts); Petrovskaia, 'La formulation tripartite'.

26 Chapter 1, pp. 46–47.

27 Petrovskaia, 'La formulation tripartite'.

28 See, for instance, C. Clifford Flanigan, 'The Moving Subject: Medieval Liturgical Processions in Semiotic and Cultural Perspective', in *Moving Subjects: Processional Performance in the Middle Ages and the Renaissance*, ed. by Kathleen Ashley and Wim Hüsken (Amsterdam: Rodopi, 2001), pp. 35–52, esp. pp. 46–47; and Teofilo F. Ruiz, *Spanish Society 1400–1600* (London and New York: Routledge, 2014), p. 147. The concept of Eurocentrism, important though it is to the study of Europe as a generalised idea, is not explored further in this book because it does not refer directly to the geographical construct, which is the focus of the present investigation. To quote Ella Shohat and Robert Stam, 'Eurocentrism does not refer to Europe in its literal sense as a continent or a geopolitical unit but rather to the perception of Europe (and its extensions

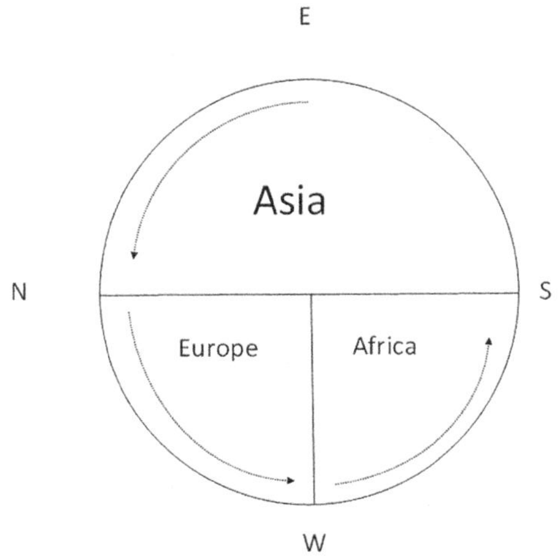

Figure 5: Narrative Progression of *Imago mundi* mapped onto a T-O schema

somewhat related to the second, is that the readers were intended to see the text in front of them, as their place in the world, as the culmination of the process of *translatio studii et imperii*, the historical progression of empire and scholarly learning westward.[29]

Now that we have established some preliminaries, let us look at what happens to this global order of description in some of the *Imago mundi* adaptations. In the Latin text the order of description is Asia–Europe–Africa, which we might depict as a counter-clockwise movement around the T-O *mappa mundi* (Figure 5).[30] Rewritings such as the *Image du monde* introduce significant changes to the text but nevertheless retain the order in which the parts of the world are discussed. The references to 'traversing' parts of the world, which in the Latin are only made in the transition between the three parts of the world, are here more frequent.[31] The structure of the description in the French text, which uses directional

around the world) as normative'; Shohat and Stam, *Race in Translation: Culture Wars around the Postcolonial Atlantic* (New York and London: NYU Press, 2012), p. 61. Shohat and Stam coin the term 'Eurotropism' to descibe the orientation towards Europe 'as an ideal Platonic Sun' (p. 62).

29 For a brief introduction to the *translatio* concept and further references, Chapter 2, p. 55.

30 See p. 47.

31 For the Latin passages, see the quotations and discussion on pp. 108, 125; the French text's trajectory through the world and technique of description is discussed in Chantal Connochie-Bourgne, 'Limites et diversités de l'Europe: le parti pris par Gossouin de Metz (*Image du monde*, 1245)', in *De la Chrétienté à l'Europe: Actes du Colloque Orléans, mai 1993*, ed. by Bernard Ribémont (Orleans: Paradigme, 1995), pp. 49–62 (p. 50).

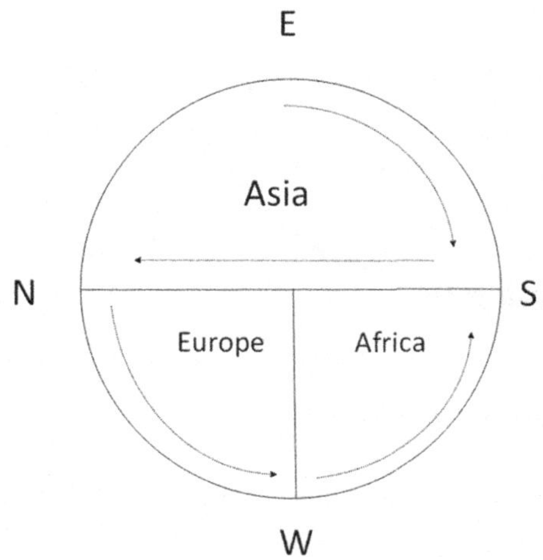

Figure 6: Narrative Progression of *Image du monde* mapped onto a T-O schema

terms ('right', 'left', 'up', 'down') invites visualisation in the form of a T-O *mappa mundi* (see Figure 6).[32] Other translations and adaptations of *Imago mundi* follow this trajectory through the inhabited world consistently, with variation limited primarily to the way the reader is guided throguh the map/text by means of transitional passages, directional terms, and so on, as well as the order of the regions within the parts of the world, or of locations within regions.[33]

The counter-clockwise direction of the original text's hodoeporical description is worth commenting on.[34] Although I describe the *Imago mundi* trajectory along the map from the viewer's perspective (looking at the map from the outside) as 'counter-clockwise', the intention in the hodoeporical description is probably that the reader is looking at the world from the starting point in the east, and imagining proceeding along the description by turning right towards the north. In medieval terms, one follows the counter-clockwise direction by turning right when facing the object of the

32 Figure 6 uses Connochie-Bourgne's summary of the direction taken by the text; 'Limites et diversités', p. 50. The trajectory, when mapped out schematically, follows a criss-crossing pattern rather than the outlines of the map, and is reminiscent of the movement of a hand making the sign of the cross.

33 See the discussion of changes made to transitional passages in the *Semeiança del mundo*, pp. 130–32.

34 The term is anachronistic (it post-dates the invention of the clockface) but is used here for convenience.

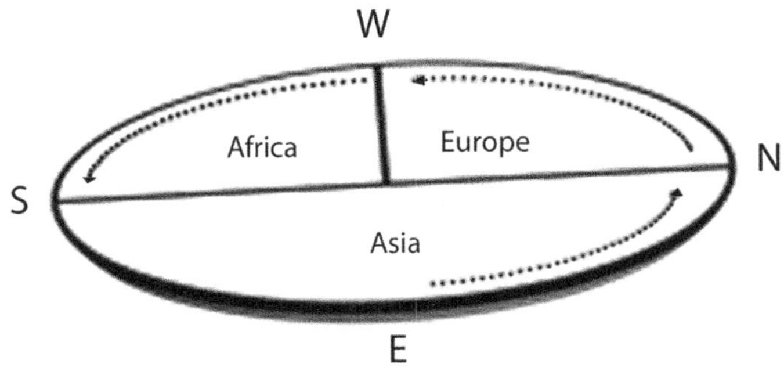

Figure 7 The direction viewed from the perspective of reader inside the map

circuit.[35] The privileging of this direction for religious processions, and thus probably also for the reading of the map in our texts, seems to be connected to the privileging of right over left in the Western tradition.[36] In the case of the trajectories described here, we should take the movement as being performed by the reader, who starts in the east and proceeds to the right (in a northward direction) along the T-O map following a procession-like circuit around the world (Figure 7).

It is striking that in some of the rewritings of the *Imago mundi* this hodoeporical order is altered extensively. Fazio degli Uberti's fourteenth-century geographical vision-poem *Il Dittamondo* is an example of this. Interestingly, being a vision poem in the style of Dante it is not merely hodoeporically constructed, but has the narrator literally travel, accompanied by a succession of guides. It is also particularly interesting for the purposes of this discussion because it is a text whose different order of description has been used to argue for independence of the *Imago mundi* tradition. It must be noted that a full study of geographical sources still needs to be undertaken

35 Given lack of clockfaces as a reference point, this is also sometimes described as 'sunwise', since it follows the trajectory of the sun in the Northern hemisphere (the movement is performed by the sun, with the observer looking east; this also explains the term for this direction, which survives from medieval Ireland, *dessel/deiseal*). This is used in describing saints' processions in a number of *Vitae*; see Tomás Ó Carragáin, *Churches in Early Medieval Ireland. Architecture, Ritual and Memory* (New Haven: Yale University Press, 2010), p. 311. Confusingly, *deiseal*, or sunwise course, is often described as clockwise (presumably because when facing the clock the observer sees its hand move towards the right); *Electronic Dictionary of the Irish Language*, s.v. *dessel*; https://dil.ie/15791 (accessed 2 November 2024).
36 See Robert Couzin, *Right and Left in Early Christian and Medieval Art* (Leiden and Boston: Brill, 2021), esp. pp. 5–7. Yi-Fu Tuan suggests that the feature is global; see Tuan, *Space and Place: The Perspective of Experience* (Minneapolis and London: University of Minnesota Press, 1977), pp. 43–44.

for this text, and the arguments here relate exclusively to its use of the *Imago mundi*, which is just one of degli Uberti's sources.[37]

Two previous suggestions open up the possibility of a direct connection between this text and Honorius's encyclopedia. In the first place, Nadia Belliato has pointed out that despite degli Uberti's insistant citation of classical authorities, his sources were probably primarily medieval.[38] More direct is Marie-Odile Garrigues's suggestion that, based on the similarity of place-name forms and names of peoples, including those which in Honorius's text are 'erroneous', the *Imago mundi* must have been one of the *Dittamondo*'s sources.[39] It is also probable that Fazio degli Uberti indeed used Brunetto Latini's *Tesoretto* and *Trésor* as sources of information, and if so the *Dittamondo* belongs in any case to the *Imago mundi* tradition.[40]

Even if Honorius's work was merely one of many texts degli Uberti used, the *Dittamondo* presents an interesting link between the *Imago mundi* tradition and a much wider network of texts, because it also had a wide circulation and influenced others. The most notable example is Matteo Maria Boiardo's late-fifteenth-century romance *L'inamoramento de Orlando*, also known as *l'Orlando Innamorato* ('Orlando in love'), to which Ariosto's *Orlando Furioso* quoted at the beginning of this book is a sequel.[41] In the

37 Johannes Bartuschat, 'La forma allegorica del "Tesoretto" e il "Dittamondo" di Fazio degli Uberti', in *A scuola con Ser Brunetto: la ricezione di Brunetto Latini dal Medioevo al Rinascimento*, ed. by I. Maffia Scariati (Florence: SIMSEL–Galluzzo, 2008), pp. 417–35 (p. 425).

38 Nadia Belliato, 'Per un commento al "Dittamondo": il Paradiso terrestre e la personificazione di Roma (i.xi)', in *Quaderno di italianistica 2015*, Sezione di Italiano dell'Università di Losana (Pisa: ETS, 2015), pp. 33–48 (p. 34).

39 Marie-Odile Garrigues, 'Honorius Augustodunensis et l'Italie', *Mélanges de l'École Française de Rome. Moyen Âge, Temps Modernes*, 84 (1952), 511–30 (pp. 525–26). Garrigues cites the unpublished study by Mario Nerva, 'L'"Imago mundi" di Onorio Augustodunensis e le scienze naturali nel dodicesimo secolo' (unpublished MA thesis, Université de Montréal, 1971), pp. 162–64, stating that it contains an overview of *Imago Mundi*–derived works. I was unable to consult this thesis due to library policy of the Université de Montréal.

40 Belliato, 'Per un commento', pp. 35–36; Garrigues, 'Honorius Augustodunensis et l'Italie', pp. 525–26. See also Bartuschat, 'La forma allegorica', and the discussion of Latini's *Trésor*, below. The issue of multilingualism also deserves a mention here. The *Dittamondo*, inspired by Brunetto's poems, contains sections in Old French, Occitan, and in Greek; Bartuschat, 'La forma allegorica', p. 426.

41 For a discussion, with further references, see Andrea Canova, 'Altri furti boiardeschi ("Inamoramento de Orlando", II, xxviii)', *Parole rubate / Purloined Letters*, 18 (2018), 163–72 (esp. pp. 166–69), available online at www.parolerubate.unipr.it (accessed 11 April 2024). Canova refers in particular to the following passages in the *Inamoramento* which appear to quote *mirabilia* from the *Dittamondo*: M. M. Boiardo, *L'inamoramento de Orlando*, ed. by Antonia Tissoni Benvenuti and Christina Montagnani, 2 vols (Milan and Naples: Ricciardi, 1999), II, p. 829 (II, i, 57), p. 1267 (II, xviii, 34), p. 1351 (II, xxii, 11), pp. 1508–09 (II, xxviii, 29); cited by Canova, p. 169 n. 12. For

following I contribute to the discussion of degli Uberti's use of Honorius by showing that the *Dittamondo* only seems to diverge from the *Imago mundi* itinerary. I argue that at its base is the same hodoeporical design, and that traces of the *Imago mundi* trajectory can still be seen when the text doubles back on itself at the end of each section.

The structure of the *Dittamondo* description of the tripartite world is in the order Asia–Africa–Europe. As I observed in Chapter 1, this is the historical rather than the geographical order of regions.[42] It is worth looking closely at the apparent difference in the order of description between the *Dittamondo* and the *Imago mundi*. An important feature of degli Uberti's text is that in its counter-clockwise trajectory it doubles back at the end of each region to follow his Asia–Africa–Europe pattern. For instance, at the conclusion of the chapter on Asia (after which the text move to Africa) there is no reference to Africa, the next region in the order of the narrative, but only Europe: 'Questo ultimo paese, ch'io ti conto, / tanto si chiude in vèr settentrione, / ch'a le palú Meotide l'affronto, / lá dove Europa i suoi termini pone' ('This last land, which I recount to you, / thus it closes towards the north, / which has the *Meotides paludes* against it / there where Europe places its boundaries').[43] Picking up at this point, his following chapter has to introduce an explicit break (*Qui lascio Europa*, 'Here I leave Europe'), to move to Africa.[44] This has every appearnce of extracting the section on Europe which should have been present here (as in the *Imago mundi*), and moving it to a different position within the text, with only traces of

comments on the wide manuscript circulation of the text, see David Paniagua, '"Soccorri me, che solo non so ire": Solino in aiuto di Fazio degli Uberti', *Cento Pagine*, 3 (2009), 10–19 (p. 13); available at https://www.openstarts.units.it/handle/10077/3222 (accessed 11 April 2024).

42 Chapter 1, p. 47. It has been suggested that the *Dittamondo* relies on Solinus's *Collectanea rerum memorabili* (also known as *De mirabilibus mundi* or *Polyhystor*). However, the order of the regions in Solinus is different again: Europe–Africa–Asia. The reference here is to the description in Book I, chapters 8–10: Fazio degli Uberti, *Il Dittamondo e le rime*, ed. by Giuseppe Corsi, 2 vols (Bari: Giuseppe Laterza & Figli, 1952), I, pp. 23–33. The parts of the world are subsequently visited in the order Europe–Africa–Asia (in Books I-VI), but the descriptions there are of historical and allegorical rather than purely geographical nature. Solinus, functions as Fazio's geographical guide on this journey, in imitation of Virgil's role in Dante. See Paget Toynbee, 'Brunetto Latino's Obligations to Solinus', *Romania*, 89 (1894), 62–77 (esp. p. 63 n. 1) for the reliance on Solinus's *Collectanea* in the *Dittamondo*. See also Paniagua, '"Soccorri me"', esp. p. 16. There is an earlier encounter with Ptolemy in the text; for a comparison with Brunetto Latini's Ptolemy and Dante's Virgil, see Bartuschat 'La forma allegorica', pp. 423–24, 430–33, and p. 432 n. 25. On Solinus's function in the *Dittamondo*, see Paniagua, '"Soccorri me"'.

43 Fazio degli Uberti, *Il Dittamondo*, ed. Corsi, I, p. 26.

44 Fazio degli Uberti, *Il Dittamondo*, ed. Corsi, I, p. 27.

the original design remaining as telltale signs at the end of the Asia and beginning of the Africa section.[45]

The break is particularly striking given the author's use of hodoeporical terminology. The term *lasciare* ('to leave'), used of places as well as objects and applied here to the conceptual movement across the textual map, is the same as that used for physical movement across space, as, for instance, in Ruggiero's journey described in the quotation from *Orlando Furioso* with which my book opened: 'lasciato avea [...] tutta l'Europa' ('He had left the European mainland').[46] The beginning of *Dittamondo* Chapter X, dedicated to Europe, shows a similar slight of hand (and similarly hodoeporical terminology): 'ritornar voglio in vèr settentrione, / dove lassai Europa in oriente' ('I wish to return towards the north / where I left Europe in the east').[47] The audience of degli Uberti's text, as that of the *Imago mundi* quite literally travels the world.

Within Europe, degli Uberti follows the same counter-clockwise trajectory as we have seen on the macro-level. He starts with Germany, moving northwest, through Britain and Ireland, south towards France, then through Spain, Italy, and finally Greece.[48] This is a slightly simplified account, as there are lists within each region which complicate this itinerary. It can be seen from this summary, however, that the order is not that of *Imago mundi*, though it follows, on the level of the regions, the counter-clockwise directionality of Honorius's overall design. The exact nature of the relationship between the *Dittamondo* and the *Imago mundi* tradition, therefore, cannot be easily solved and must await its own dedicated study. For now, it can be tentatively concluded that in the hodoeporical trajectory, and the base order of the regions which can still be seen in trace form at ends and beginnigs of its major sections, the *Dittamondo* echoes Honorius's encyclopedia. In terms of its general purport, too, it would be typical of the *Imago mundi* tradition as a compilation of established knowledge based on earlier sources and presented in a more accessible, simplified, form. While not in itself conclusive, this offers some additional supporting evidence for Belliato's and Garrigues's suggestions mentioned above. The French text also brings us to our next

45 By contrast, Solinus ends his description of Europe with Spain and then moves to Lybia (Chapters XXIII–XXIV); *Collectanea rerum memorabilium*, ed. by Th. Mommsen (Berlin: Weidmann, 1895), p. 107.

46 Ludovico Ariosto, *Orlando Furioso: A Selection*, ed. by Pamela Waley (Manchester: Manchester University Press, 1975), p. 55; trans. by Guido Waldman, Oxford World Classics (Oxford: Oxford University Press, 1974), p. 52; quoted in this book, pp. 13–14.

47 Fazio degli Uberti, *Il Dittamondo*, ed. Corsi, I, p. 27.

48 Fazio degli Uberti, *Il Dittamondo*, ed. Corsi, I, pp. 27–33.

point: variations in the hodoeporical description between different versions of the same text.

Significant Itinerary Variations in the Latin Text

In the *Imago mundi*, while the order of the tripartite (or quadripartite) design remains stable, there is variation in the order of hodoeporical description within the regions between the various versions of the Latin text. The most interesting, and most radical, among these examples of variation is in the Italy chapter (I.26).[49] Flint has described this as a 'tangle' and has observed that the consistency in transposition of specific sections of the description accross the manuscript tradition is such as to suggest that this is more than a scribal error.[50] I focus on this chapter here because it both offers a fascinating glimpse into the ways in which the original text's flexibility forms the later tradition, and enables us to determine exact relationships between branches of that tradition. The variation has proved instrumental in identifying the different versions of *Imago mundi* used in the Welsh translations (*Delw y Byd*), since it is inherited by those texts.[51]

Of the two Welsh translations, the text known as Version B, is attested in the 'White Book of Rhydderch' Aberystwyth, National Library of Wales, Peniarth 4–5 (s. XIV), and in the 'Red Book of Hergest' Oxford, Jesus College, MS 111 (s. XIV$^{ex.}$), col. 510 onwards.[52] Although it abridges considerably, and omits one of the sections of the passage, the order of the Italian passage in this version agrees with the Latin text as it is printed in Flint's edition, which corresponds to the later versions of the *Imago mundi*. Meanwhile, *Delw y Byd* Version A, preserved in the Red Book col. 987 onwards, follows a different order of the passage, found in an earlier recention of *Imago mundi*. This variation has made it possible to identify the manuscript of

49 Quoted in full below, pp. 146–47.

50 Flint, ed., *Imago mundi*, p. 14. We have seen a similar argument applied to the French text's transposition of parts of Europe to Africa in Chapter 4.

51 Natalia I. Petrovskaia, '*Delw y byd*: une traduction médiévale en gallois d'une encyclopédie latine et la création d'un traité géographique', *Études celtiques*, 39 (2013), 257–77, (pp. 267–70); Petrovskaia, ed., *Delw y Byd : A Medieval Welsh Encyclopedia*, MHRA Library of Medieval Welsh Literature (London: MHRA, 2020), pp. 7–8 n. 29; Flint, ed., *Imago mundi*, pp. 61–62. By contrast, the German translation, for instance (which, like the Welsh, translates the Latin faithfully), follows the order of the passage as it is in Flint's edition; *Das puch von der Forme und pildnuss der Welt* in Germ. Fol. 595, f. 86r l. 31–86v l. 8.

52 I specify column number for the Red Book because this manuscript contains both versions A and B of *Delw y Byd*; Petrovskaia, ed., *Delw y Byd*, pp. 8–9.

the Latin text closest to the Welsh translations, namely Oxford, Bodleian Library Rawlinson B.484 (s. XII$^{ex.}$).[53] This manuscript has been grouped by Flint with others containing the 1110 redaction. However, on the basis on this difference in the Italy chapter it appears that this group should be further subdivided, allowing for a more accurate understanding of the early transmission of the Latin text in England and proving once more its original flexibility and pliability. Here, I propose to use the hodoeporical logic of the descriptions to show that the Rawlinson B.484 reading probably reflects an intermediate stage in the design of the Latin text. In order to facilitate analysis, I label and discuss the components of the hodoeporical description first, followed by the annotated passage.

In her edition, based on the the post-1123 version of the text, Flint labelled the sections susceptible to re-ordering consecutively with numbers (1) through (4), and for the sake of consistency I keep her numbering.[54] There are two further variants on this order, both belonging to manuscripts ascribed by Flint to the text's earliest, 1110 version. The order preserved in manuscripts of the 1110 version proper, exemplified by Cambridge Corpus Christi College MS 66 (s. XII$^{ex.}$, Sawley, Lancashire), is as follows: (1) *Antiqui ciuitates...*, (4) *Renus ab Alpibus nascitur...*, (3) *Padus qui et Eridanus* and (2) *Est in Italia Tuscia prouintia*.[55] For convenience, I label this the 'Corpus sequence'. The order of the alternative variant, also assigned to the 1110 version by Flint, is that preserved in the Rawlinson B 484 manuscript and the Welsh version that follows it. This is: (3) *Padus qui et Eridanus...*, (2) *Est in Italia Tuscia prouintia*, (1) *Antiqui ciuitates...*, (4) *Renus ab Alpibus nascitur....* I label this the 'Rawlinson sequence'. The difference between the Rawlinson and Corpus sequences amounts to the first two components swapping places with the last two: 3+2 and 1+4 are always together (see Table 1).

Table 1: The order of Chapter 27 sections in the *Imago mundi*

Rawlinson B 484	Corpus 66	Later versions (Flint's edn)
3-2-1-4	1-4-3-2	1-2-3-4

The Rawlinson sequence is to my knowledge unique among the Latin manuscripts. The Corpus sequence, by contrast, is attested in at least one

53 Petrovskaia, '*Delwy Byd*: une traduction', pp. 269, 272.

54 Flint, ed., *Imago mundi*, pp. 61–62.

55 The full passage, in the order in which it appear in this verion, is also reproduced below.

further manuscript: London, Lambeth Palace MS 371 (Reading Abbey, s. XIII).[56] The passage, as it appears in the Corpus and Lambeth manuscripts (f. 12r), is reproduced below.[57] The numbers assigned to the sections of the passage in the discussion above have been added for ease of reference.

> In hac est urbs Roma, a Rolumo constructa, et sic nominata. (1) Antiqui ciuitates secundum praecipuas feras ob significationem formabant. Vnde Roma formam leonis habet, qui ceteris bestiis quasi rex preest. Huius capud est urbs a Romulo constructa, laterali uero edificio utrobique disposita. Vnde et Lateranis dicitur. Brundusium autem formam cerui, Kartago bouis, Troia equi figuram habuit. (4) Rhenus ab Alpibus nascitur, et contra auilonem uergens sinu occeani excipitur. (3) Padus, et Heridanus Italie fluuius ab apenninis montibus oritur, ac mari inmergitur. Yenetia a Beneto rege prius Benetia dicta, deinde Uenetia. (2) Est in Italia Tuscia prouincia[58] a thure et sacrificiis dicta. Est et Campania a Capua ciuitate dicta, a Capi rege constructa. Ibi est et Vmbria,[59] uel Ibria inde dicta quod imbribus tempore diluuii superfuit. Est et Etrutia ab Ethrusco rege dicta. Est et Longobardia a longobardi[60] uocata. Gallia a candore populi dicitur. Gala enim Grece 'lac' dicitur Latine.

> In that is the city of Rome, consctructed by Romulus, and so called. (1) The ancients shaped cities following important beasts according to their significance. Thus Rome has the form of a lion, which rules other beasts like a king. Its head is the city constructed by Romulus, with buildings placed on either side [= laterally]. For that reason it is also called Lateran. Brindisi, indeed, has the form of a stag, Carthage that of a bull, Troy of a horse. (4) The Rhine is born in the Alps, descending in the direction of the

56 Discussed in Bettina Schöller, 'Transfer of Knowledge: *Mappae Mundi* Between Texts and Images', *Peregrinations: Journal of Medieval Art and Architecture*, 4 (2013), 42–55; the passage under discussion here can be seen on the photograph of the manuscript reproduced as Figure 3 of the article (p. 49) and is also available on the Lambeth Palace Library website at https://images.lambethpalacelibrary.org.uk/luna/servlet/s/x82y4t (accessed 1 October 2023).

57 The passage is reproduced on the basis of a transcription of the Lambeth Palace text, reproduced with the kind permission of the Lambeth Palace Library. I have expanded abbreviations, added capitalisation, and modernised the punctuation. For a published transcription of the Corpus text, see Petrovskaia, '*Delw y Byd*: une traduction', p. 267. There is minor variation between the two manuscripts, amounting primarily to spelling (e.g. *Karthago* for *Kartago*) or word choice (e.g. *appellatvr* or *dicitur latine* for *dicitur*). The translation is my own.

58 The Corpus MS has an error here, and reads *pronuntia*.

59 *Apulia* in the Corpus MS.

60 *Om.* in the Corpus MS.

> north, flowing into the bay of the ocean. (3) The Po, also called Eridanus, a river of Italy, comes from the Appenines and merges into the sea. Venice was first called Benetia from King Benetus, thereafter called Venetia. (2) There is in Italy the province Tuscany, named after incense [= *thus*] and sacrifices. There is also Campania, called after the city of Capua, built by King Capys. There are also Umbria also Imbria, called so because it survived the rains [= *imbribus*] of the time of the Flood. There is also Ethruria, called after King Ethrusc. There is also Longobardia named after long-beards [= *Lombards*]. Gallia is named after the whiteness of the people. For 'Gala' in Greek means 'milk' in Latin.

An analysis of the structure of the passage and the changes in the order in which the region is described suggests a redesign of the narrative itinerary. A structural logic can be found in each of the variants, and it is possible to see the intentionality behind the re-ordering. Figure 8 maps the four sections of the passage onto a schematic image of Italy. Section (2), beginning with the reference to Tuscany, is also broken down into the various geographical references contained within it as it contains a vividly marked internal trajectory.

Figure 8 shows that the positions of the place-names heading each section orchestrate an overall northward movement, from Rome, through Tuscany, to the River Po, and beyond the Alps to the Rhine (outside Italy): sequence (1)–(2)–(3)–(4). The subsections (2.1)–(2.4) interrupt this northward trajectory with a counter-clockwise perambulation of the peninsula – or, from the perspective of the traveller, rightward. The whole reveals a complex, intricately structured design. The arrows in Figure 8 demonstrate the trajectory. The circuit echoes the overall counter-clockwise direction of the progression through the world in the *Imago mundi*.

If we look at the Corpus sequence, (1)–(4)–(3)–(2), in the Rhine in the middle of the sequence, it introduced what might appear to be an anomaly, if we assume that Flint's later manuscripts represent the 'proper' base sequence. This seeming anomaly, however, is susceptible to a ready explanation. It is possible that in the Corpus sequence, (4) *Rhenus* may have been thought of not as the Rhine but as Reno, the river flowing through Tuscany and Emilia Romagna, and a tributary of the Po. While Reno originates not in the Alps but in the Apennines, it does curve northwards part of its way, and can be described as *contra aquilonem uergens*. In that case, although there would be no northwards progression of the sections (since Po is further north than is Tuscany), the grouping Reno–Po would be in tune with the thematic trend in the passage. In that case the logic of the passage would not be

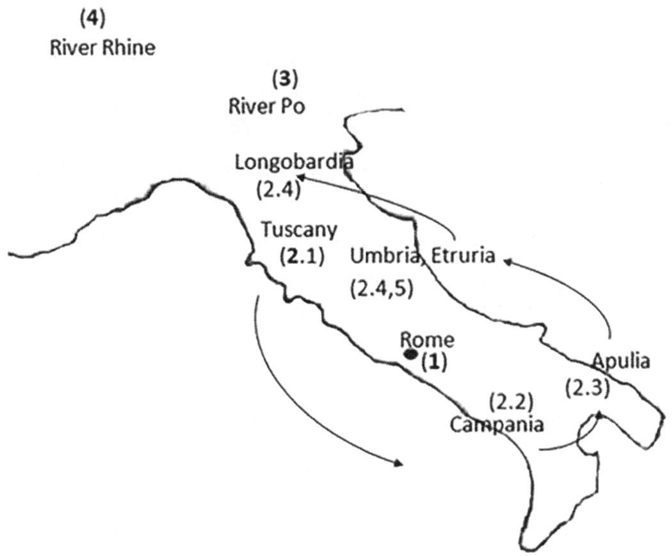

Figure 8: Map of Italy showing the position of Sections (1)–(4) of *Imago mundi* I.26

hodoeporical but rather seen in terms of topographical features, to which each of the numbered section corresponds: (1) cities, (4) and (3) rivers, (2) regions. Within each section, multiple geographical locations are included:

(1) Rome, Brindisi (+Carthage and Troy outside Italy);
(4) Reno and (3) Padus;
(2) This section preforms a counter-clockwise circuit of Italy, as in all versions of *Imago mundi*. Venice is the first item in this section, followed by Tuscany, Campania, Apulia, Umbria, Lombardy. (Since this sequence occurs in the 1110 version, it is probable that this was the original design of the passage.)

From Lombardy the text moves on to Gallia and Chapter 27. Here, the Corpus text agrees with that of the later versions, in Flint's edition: 'A flumine Reno est Gallia Belgica, a ciuitate Belgis dicta' ('From River Reno [=Rhine] is Gallia Belgica [=Belgic Gaul], named after the city Belgis').[61]

My argument here is that despite the apparent confusion between the Reno and the Rhine at the beginning of the chapter, it is the Corpus text, where the reference to Lombardy is followed by the reference to Gaul, that

61 The only difference is *Rheno* for *Reno*; Petrovskaia, '*Delw y Byd*: une traduction', p. 267; Flint, ed., *Imago mundi*, p. 62.

contains the original composition. The description of Europe in this version is organised thematically, rather than hodoeporically (as an itinerary): the link between *Longobardia* and *Gallia* is not location, but etymology. Both place-names are derived in the text through characteristics of their inhabitants: the long beards of the Longobards, and the milky pale skin of the Gauls.[62] The Isidoran inspiration is more visible in this earlier version. The changes enacted in the later versions seem to move away from this inspiration with its emphasis on etymology, and towards spatial links and the emphasis on hodoeporical organisation. One can think of this in terms of moving away from Isidore and towards *mappae mundi*. This appears to be part of the process that leads ultimately to the itinerary-like description of adaptations such as *Image du monde*, with its use of directional teminology guiding the audience's mental perambulation.

To illustrate this point, it is worth spending a little time on the Rawlinson sequence. Here, the Po and *Rhenus* are separated: (3) *Padus* + (2) *Tuscany* (and the circuit of Italy) are brought to the start of the passage, while (1) *Antiqui civitates* and the (4) *Rhenus* sections are left to the end. The Appenines, from which (3) states the Po descends, had been mentioned already at the end of I.25, *De Graecia*, and thus bringing the section concerning the Po closer to the beginning of the chapter creates a hodoeporical link to the preceeding itinerary. The Po–Tuscany link is geographical (Figure 8), while section (1) on the cities, headed with the reference to Rome, is an inserted vignette within the counter-clockwise sequence of (2). Whether Rhenus is, in this version, intepreted as the Reno or the Rhine is uncertain, but given it is found at the end of this sequence, and leads to the *Gallia Belgica* reference, it seems likely that in this case *Rhenus* is identified as the Rhine.

The changes Honorius made to the later versions yield an increasingly hodoeporical structure, using the Rhine as a connection between the discussion of Italy and the description of the various Gauls (Gallia Belgica, Francia, Ludgunensis Gallia, Narbonensis Gallia) in I.27. Following the Rhine from the Alps, to the north, permits the author to continue the narrative north to south through Gaul, leading the reader via Aquitaine at the end of I.27 to Spain in I.28.

The fact that Gallia is already mentioned after the reference to Venice and before the reference to the Rhine, thus creating a doublet, echoes the double refernce to the Appenines (in I.26 and I.27) and the double reference to the Rhine itself, discussed in relation to the Rawlinson sequence above. Two observations can be made here. In the first place, this doubling represents the same phenomenon as the overlapping fuzzy sets of the *Image du monde*

62 Flint ed., *Imago mundi*, pp. 61–62.

discussed in Chapter 4. This reinforces my suggestion that Jostkleigrewe's reading of the intentional overlap in *Image du monde* is applicable to other members of the wider tradition, and that the French text is not actually anomalous, as Jostkleigrewe suggested. The second observation is that if we see this repetition of place-names in the hodoeporical description as overlaps between fuzzy sets, that explains why it is often difficult to see boundaries in the *Imago mundi* and its adaptations. The text moves through the world, from one region to another, but in doing so names the areas through which it passes (place-names, cities), rather than lines or boundaries that it might cross.[63]

Such a way of reading the textual map might also throw some light on the absence of borders on medieval *mappae mundi*. This can be seen as a function of the difference between medieval and modern maps in terms of purpose. The medieval map, whether visual or textual, represents the physical aspect of the creation, synchronously, with reference to past moments (Jostkleigrewe's historico-geographical model), through which the viewer travels, while the modern map often has a geopolitical function, and thus shows borders of political unities.

The Latin text's description of Europe and its regions, as demonstrated in the discussion of its Italy chapter and its itinerary style, is flexible and susceptible to re-organisation and rewriting, which do no more violence to the information contained in the text than a car GPS would to the road network schema when mapping an alternative route for the user.[64] The change to the organisation principles of the text in the description of Italy also shows that the *Imago mundi* (before one even approaches its vernacular adaptations) is a much more fluid text than had hitherto been assumed.[65] Once such a complete change in the organising principles of the text is observed in the Latin text itself, it becomes easier to see how the text was used and adapted in vernacular compositions where the organisational principles of the geographic description differ starkly from what might be expected of a faithful translation of the *Imago mundi*.

63 See the discussion of boundaries above, pp. 109–11, 120–22.

64 For a discussion of a medieval map, demonstrating that it was constructed on the principles of an itinerary, see Thomas O'Loughlin, 'An Early Thirteenth-Century Map in Dublin: A Window into the World of Giraldus Cambrensis', *Imago Mundi*, 51 (1999), 24–39, esp. pp. 28–31.

65 In her discussion of the differences between the versions of the *Imago mundi*, Flint refers to 'omission', 'expansion', and 'emendation', but does not give examples of the latter; *Imago mundi*, pp. 36–39. Confusingly, 'omission' in her discussion is often used in reference to the shorter earlier versions.

Local Interest

Since the main change to descriptions of parts of Europe in adaptations of the *Imago mundi* usually concerns those areas where the adaptation in question was produced, it is worthwhile examining what happens to the already flexible description of Italy in adaptations produced there. A useful starting point is provided by Brunetto Latini.

In Chapter CXXIII of Latini's *Trésor*, dedicated to Europe, Italy occupies a dominant part, taking up an amount of space proportional to the rest of the chapter.[66] The description of Italy commences by outlining its continental edge, west to east (it refers to the Alps being where Italy *finische*, 'ends'), followed by reference to two of its most important rivers: Po (Brunetto also gives the Latin, *Padus*), and *Rosnes* (= Rhône, for which Brunetto does not give the Latin *Rhenus* found in the *Imago mundi*).[67] The geographical course followed by Latini's description of Italy is as follows: Tuscany, Rome, Campania, Apulia, Calabria, Sicily, Romagna, Lombardy, Treviso, Sardinia and Corsica, and, facing Italy, Istria (Figure 9).

Figure 9 shows that the sequence of regions in Brunetto Latini's description of Italy follows a complex pattern. Comparison with the *Imago mundi* sequence in Figure 8 establishes that the Latin text served as the base for Latini's description, starting in the north-west, and moving south, east, and north again, but turning back west before doing so. The major difference introduced by Latini is moving Rome from its original position at the head of the sequence to a medial position dictated by hodoeporical considerations, placing it between Tuscany and Campania. In other words, this represents a further development in the hodoeporical structuring of the account. If we take the *Imago mundi* (Figure 8) section numbers and put them in the order in which we find the information in Latini, we get: (3)–(4)–(2.1)–(1)–(2.2)–(2.3)–[new additions]–(2.4)–[new additions]. The main interruptions in this sequence in Latini's text are constituted by the insertion of Calabria, Sicily, Corsica, and Sardinia into the sequence, which create a westward move before and after the return north, resulting in an 8-shaped trajectory. This is caused by a fundamental departure from the

66 To illustrate the point, we might consider the proportions based on Carmody's edition; Brunetto Latini, *Li Livres dou Tresor*, ed. by Francis Carmody (Berkeley: University of California Press, 1948). Following the introduction of the region of Europe, which occupies just under four lines of the printed text, the discussion of Italy takes up 70 lines; Brunetto Latini, *Tresor*, ed. Carmody, pp. 115–17. The rest of Europe, including Greece, Germany, France, Spain, and Britain, occupy the remaining 100 lines (pp. 117–19).

67 Brunetto Latini, *Tresor*, ed. Carmody, p. 115.

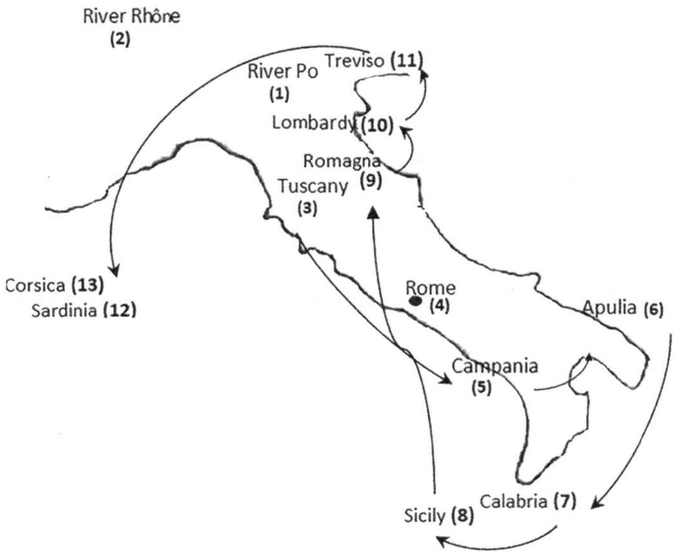

Figure 9: Map of Italy showing the order of description of its regions in the *Trésor*

basic structure of the Latin text, which – we must remember – had separated all Mediterranean Islands from its description of the mainland and had them as a separate fuzzy set.[68] Latini puts them back into Europe. Isolating all islands together into a separate section of the text, as in the *Imago mundi*, suited neither the purpose of Latini's new encyclopedia nor the new central focus that he introduced into his description of Italy.

It is tempting to read this trajectory in the context of the *Trésor*'s overriding obsession with ecclesiastical territorial structures (discussed in Chapter 3). While I would not wish to press the point too far, the 8-shaped trajectory through Italy resembles the movement of the hand making the sign of the cross, as if the territory is being blessed by the progress of the description. This intepretation of the trajectory is hightly tentative, but given Latini's ecclesiastical focus in this text, perhaps not altogether fanciful.[69]

Although, as has already been discussed above, medieval maps and geographical descriptions do not, as a rule, accord the same stress to borders and boundaries as modern descriptions do, the texts examined

68 See p. 121–27.

69 This special treatment of Italy in preference to other regions is not without parallel. One thinks in particular of the privileged depiction of Italy on the mural maps of the Vatican Terza Loggia, where it is one of only two regions in the world depicted entirely in gold. The other is the Holy Land. See discussion in Florio Bamfi, 'The Cosmographic Loggia of the Vatican Palace', *Imago mundi*, 9 (1952), 23–34 (p. 25).

in this chapter display an awareness of the distinction between different regions and areas. This awareness appears to be articulated primarily by means of hodoeporical description where different areas form re-orderable components in the itinerary across the world. The manipulation of this hodoeporical order, made possible by the flexible structure of the original text, appears to reflect regional interests and the context of production of each adaptation. The re-ordering of regions discussed is usually not the only change implemented and often goes hand-in-hand with other alterations, as with Latini's switch from the Roman imperial terminology of the *Imago mundi* to ecclesiastical terminology.

All texts examined here, apart from the fourteenth-century German translation of *Imago mundi*, demonstrate predilections towards particular regions through the process of rewriting the description to fit the home area of the text and its audience, which is likely to be connected to senses of regional identity.

Conclusion: *Looking Back and Looking Ahead*

Abstract: The conclusion reflects on the main arguments of the book and the reasons found for the *Imago mundi*'s enormous success in the medieval period. This success is due to two main factors: the fact that the text was already archaic and thus by definition could not become outdated, and the fact that the geographical areas described constitute what we can now define as fuzzy sets. The lack of definite boundaries allowed subsequent translators and adaptors of the text to superimpose their own maps of Europe onto the geographical template provided by Honorius's treatise. The conclusion also looks forward in time to the continued preservation and transmission of the text and its adaptations in incunabula and early modern printed books.

Keywords: fuzzy framework; retargeting; refocusing; *Imago mundi* tradition; early printed books

Transforming Europe is about conceptualised space. Its focus is on how the medieval geographical texts of the *Imago mundi* tradition described Europe. My main argument is that two factors, archaism and fuzziness, contributed to the enormous success (measured by its spread and amount of texts that use it) of this tradition, based on *Imago mundi* Book I.[1] The archaism factor is that the textual map provided by Honorius remained in use for centuries and did not go out of date because it was out of date to start with. The fuzziness factor is that this textual map was adaptable to different regions because it did not assign regions a stable and fixed position within its spatial framework. All of the regions described belong to (and themselves are) what we can now describe in technical terms as fuzzy sets.

1 Other factors include the simplicity of Honorius's language, for example.

Petrovskaia, N.I. *Transforming Europe in the Images of the World, 1110–1500. Fuzzy Geographies.* Taylor & Francis Group, 2025
DOI 10.5117/9789048563166_CON

The importance of understanding the medieval geographical concept of Europe lies in its use as an analytic category. If we do not know what the term means, we cannot use it, and, as Klaus Oschema points out, risk reading into it meanings that are significant to us but that were not primary to its original users.[2] Despite the lack of a real geographical 'continent', texts, including the medieval texts discussed in this book, construct a geographical entity called 'Europe' and thus make it real.[3] These texts help the audience locate themselves in the world. Multiple narrative texts in different languages employed Honorius's fuzzy framework to frame their view of their world, providing a setting for the action. In the case of some of the adaptations, this use is subject to alterations inherent not only in the refocusing of the text on its new host region, but also in the context of the author's own repositioning (of the work and of himself). For the *Image du monde*, for instance, crucial is the text's own position within the progression of the *translatio studii et imperii* movement, as an expression of the French status as the heirs of the empire and learning of the ancient world.[4] The framework being used is both spatio-temporal and cultural. The French text thus places itself and its information (the *Imago mundi* map) into a model formed in its host culture.

A consistent feature in the adaptations examined here has been the reordering of the material to reposision the reader within the geographical framework provided. As Caroline D. Eckhardt observes, 'Whether the lands on the margins of Europe would represent themselves as distinct from the continent, or as integral to it, was a contingent rather than consistent decision, then as now.'[5] The observation has particular poignancy in the twenty-first century. Eckhardt's focus is on fourteenth-century England, but although the marginal position of England, and Britain more generally, in medieval geographical and cartographical representations of the world has attracted much attention, as Kathy Lavezzo points out, 'the English held

2 The concern is voiced by Klaus Oschema in *Bilder von Europa im Mittelalter*, Mittelalter-Forschungen, 43 (Ostfildern: Thornbecke, 2013), p. 18.

3 Compare the discussion of medieval spatial representations of France in Nathalie Bouloux, 'From Gaul to the Kingdom of France: Representations of French Space in the Geographical Texts of the Middle Ages (Twelfth–Fifteenth Centuries)', in *Space in the Medieval West. Places, Territories and Imagines Communities*, ed. by Meredith Cohen and Fanny Madeline (Farnham and Burlington: Ashgate, 2014), pp. 197–217, esp. pp. 197–98 on terminology.

4 O. H. Prior, ed., *L'Image du monde de maitre Gossuin. Rédaction en prose* (Lausanne and Paris, 1913), p. 77.

5 Caroline D. Eckhardt, 'One Third of the Earth? Europe Seen and Unseen in the Middle English Chronicles of the Fourteenth Century', *Comparative Literature*, 58.4, 'The Idea of Europe' (2006), 313–38 (p. 314).

no monopoly on geographic marginality in the West'.[6] Thus, Eckhardt's observations on the relationship of lands on the margins of 'Europe' to the perceived whole can be applied to a broader range of corpora. Caxton's addition of material relating to his own position and geographical understanding thus opens another layer of enquiry: the role of the author's (taking the term broadly to include redactors and translators) position in determining their attitude to the information conveyed by the text.[7] In terms of geographical definitions of Europe this becomes particularly important for authors and versions of the text originating from marginal areas. The concept of the fuzzy set is particularly useful for understanding the flexible and equivocal state of these marginal areas.

The alterations we find in some of these texts, such as the *Semeiança*, for instance, might seem to call into question their status as 'translations' of the *Imago mundi*, but, before pronouncing judgment, it may be useful to call to mind the distinction, offered by Umberto Eco, between two types of faithfulness in translation – faithfulness to 'not necessarily the intention of the author [...], but the *intention of the text* – the intention of the text being the outcome of an interpretative effort on the part of the reader, the critic or the translator'.[8]

In the case of the tradition examined in this book, it may be necessary to add two additional considerations, which constitute the retargeting and refocusing of the material. We have seen an example of such retargeting in Gossouin's adaptation, which retargeted the Latin encyclopedic *Imago mundi* I towards a new, lay audience, giving it a new genre label, *livre du clergie*.[9] The retargeting of the *Semeiança* is not as sharp as for the *Image du monde*, but the (relatively common) re-attribution to Isidore can be read alongside the added focus on Spain as a retargeting for a Castilian audience. In most of the rewritings, refocusing has also been a feature, recogniseable in the addition of information concerning the region local to the adaptation itself. An even more striking feature of this refocusing is found in the changes made to the order in which the regions are presented in some of these texts, leading to the repositioning of the local region as the

6 Kathy Lavezzo, *Angels on the Edge of the World: Geography, Literature, and English Community, 1000–1534* (Ithaca, NY: Cornell University Press, 2006), p. 53. Lavezzo follows this observation with a discussion of the representations of Ireland, in particular by Gerald of Wales, cited above, p. 119.

7 See above, p. 84 n. 3 and p. 112.

8 Eco, *Mouse or Rat? Translation as Negotiation* (London: Weidenfeld & Nicholson, 2003), p. 5, emphasis Eco's.

9 See Chapter 4.

last in the sequence within Europe. This has been a feature of the *Image du monde*, the *Semeiança*, and *Laberinto de Fortuna*, for instance.

It remains to be seen whether the features I have outlined for the *Imago mundi* tradition are shared by other medieval encyclopedic traditions or whether this particular construction is unique to Honorius and his direct successors. Before we let the hippogriff fly us to new and uncharted lands, however, let us look briefly over our shoulder to see where we have been.

Looking Back

Transforming Europe began with two main objectives: to introduce the incredible range and influence of the *Imago mundi* family of texts, and to provide some sense of why it had that incredible range and interest. With regard to the latter, I used the texts' changing descriptions of Europe as a focal point for the discussion. Chapter 1 set the scene by providing a brief introduction to what is currently known of the *Imago mundi* itself, its classification as 'encyclopedia', issues surrounding its authorship and attribution, composition context, and finally its contents. Like the *Imago mundi* itself, this section presented the reader with well-trodden territory, material long known and well established. This was followed in Chapter 2 by an introduction to the large family of *Imago mundi* translations and related texts. Like many large families, this one is difficult to encompass and its members sometimes have complicated interrelationships. Genre boundaries appear to be routinely crossed by this material both in transmission between languages and into new literary contexts, and within texts, adding didactic, textbook material to romance and narrative texts, and adding narrative to encyclopedic material. As with the interaction of text and illustration, the emerging pattern is one of spectacular symbiosis rather than of parallel or converging traditions. I have attempted to introduce some order into this complicated tangle of texts by grouping them into fuzzy sets. The choice to keep the sets intentionally fuzzy was made in order to provide the classification with the necessary flexibility for subsequent research to introduce alterations and shift members from one set to another, without requiring that the entire classification system be totally redrawn. Adjustments can be made to how I have classified individual texts, and more texts can be added, without the need to rebuild the system. The fuzzy sets are:

- vernacular translations of the *Imago mundi*;
- adaptations inserting translations of *Imago mundi* fragments into new encyclopedic texts;

- translations and adaptations of fragments and extracts incorporated into non-encyclopedic (narrative) texts;
- adaptations into other Latin texts.

To the first category belong – to my mind, unambiguously – *Semeiança del mundo*, *Delw y Byd*, the Italian text of the bilingual Estense manuscript, and the German *Das puch von der Forme und pildnuss der Welt*. In the second group I have placed the *Mappemonde* of Pierre de Beauvais; the *Image du monde* of Gautier/Gossouin de Metz; *La Petite Philosophie*; the *Divisiones mundi* of Perot of Garbelei/Garbelai; the *Mirrour of the World* of William Caxton; *L'Ymagine del mondo*; the *Mapa del mond* of Vivaldo Belcalzèr; and perhaps somewhat controversially, Brunetto Latini's *Trésor*. The two Hebrew versions are placed here also, but I have not been able to study them in detail and they do not feature further in this book.

The third category, which takes the material from the *Imago mundi* across genre boundaries (though these are also fuzzy) includes, in my classification: the Catalan Atlas/Ecumene Chart, which contains extracts verbatim translated into Catalan (or extracts from a verbatim Catalan translation); Jacob van Maerlant's *Alexanders geesten*, which contains a section likewise translated quite closely; Juan de Mena's *Laberinto de Fortuna*, which contains an adaptation; and the Old Norse *Fóstbræðra saga* and *Eiriks saga víðförla*. The latter two texts were not discussed further in the book, either, as they do not contain sections on Europe. To this category – with something of a question mark – also belongs the *Dittamondo* of Fazio degli Uberti. While this text has often been assumed not to have drawn on the *Imago mundi*, in the discussion of the hodoeporical structure of these texts in Chapter 5 I have argued that it belongs in this group after all.

From the final category of Latin texts that use the *Imago mundi*, I have only discussed the work of Gervase of Tilbury as it is balanced in interesting ways with that of Brunetto Latini. Further work is needed on the use of *Imago mundi* by Thomas of Cantimpré, Vincent of Beauvais, Michael Scot, Gervase of Tilbury, Godfrey of Viterbo, and John of Saint-Victor. A full study of the Latin tradition of adaptations and uses of *Imago mundi* is a major endeavour and may well prove that the current estimation of the influence of Honorius's text does not fully acknowledge its siginificance.

Where Chapters 1 and 2 served as an introduction to the breadth of the material, bringing it together for the first time, Chapters 3 and 4 provided tentative answers to the question of why this material spread so widely. Chapter 3 explored the striking archaism of Honorius's textual map. The example chosen is the chapter beginning his section on Europe, where the marginal areas of that region are described in terms which are already outdated by the twelfth century. The chapter then moved to explore the implications of this purposeful antiquarianism for the text's transmission. It illustrated somewhat different attitudes to the original displayed by its translators. Their choices seem to have been dictated by the requirements of the new text, attesting to the flexibility of the original, despite – or perhaps because of – its archaising nature. An example of the retention of archaic nomenclature is seen in the Welsh and Spanish translations, while in the works of Gervase of Tilbury and Brunetto Latini an ecclesiastical map of Christendom is overlaid on Honorius's underlying map that, depending on our attitude towards it, we might choose to call either 'anachronistic' or 'timeless and therefore flexible'.

In Chapter 4 we saw that in the text these regions are mapped – as Jostkleigrewe has shown for the *Image du monde* – according to a historico-geographical view that is not exclusively spatial. This patterning might present conceptual difficulties to a modern viewer used to a sharply delineated geometric representation of geographical space. Within the texts in question it often works alongside a hodoeporical method to the description and the world is described by taking the audience on a virtual journey through it, discussed in Chapter 5. This might also explain why it is that the regions can be fuzzy and that the borders might not matter – as the text moves through space, taking the audience on a journey of textual exploration, it matters what they find *inside* the different regions visited, not necessarily how the regions are bounded.

Following on the discussion of the why of the material's popularity in Chapters 3 and 4, the focus of Chapter 5 was how and in what context this material was presented to its audiences. Chapter 5 dealt with the issue of directed reading of the map. I have termed this the hodoeporical description technique, and have argued that the lack of boundaries works well with the itinerary-like flow of the narrative. We also explored how this itinerary changed in some of the *Imago mundi* adaptations, and why.

As Roger Ellis aptly describes it, 'Translation in the Middle Ages, as, indeed, at any time, witnesses parodically, after the manner of a distorting mirror, to frontiers geographical, cultural and linguistic, and assists routinely in their creation and dissolution'.[10] This observation is particularly true of

10 Roger Ellis, 'Translation and Frontiers in Late Medieval England: Caxton, Kempe, and Mandeville', in *Frontiers in the Middle Ages: Proceedings of the Third European Congress of Medieval Studies (Jyväskylä, 10–14 June 2003)*, ed. by O. Merisalo and P. Pahta, Textes et Etudes du Moyen

the *Imago mundi* texts, which describe the boundaries they themselves cross, and provides a useful perspective on the twofold problem presented by the *Imago mundi* tradition in its relationship to Europe: as a text that crosses cultural, linguistic, political, and genre boundaries while traversing the regions it itself describes. Ellis's use of the term 'frontier' also resonates strongly in the argument of this book: that we are dealing here not with linear borders – whether between geographical territories or between text types – but with areas of overlap between fuzzy sets.[11]

This book has aimed to outline the evolution processes of two fuzzy sets: the development of the *Imago mundi* tradition and the transformation of the geographical description of Europe specifically. I have aimed to show that Honorius's template of Europe is so malleable and so easy to use for later authors because it is already antiquated and purposefully and formally fuzzy. It yields itself to overlay with an array of multi-layered topographical, cultural and political maps, often providing varied and not necessarily overlapping definitions within the framework of the same text. As Michael Wintle observes, there is more than one type of border: geographical, cultural, and political, and these rarely coincide.[12] We might adapt this statement to say there is more than one type of spatial fuzzy set: geographical, cultural, and political, and these often partly overlap. There is also, consequently, more than one Europe.

The present study was limited to the examination of the different perceptions of the borders and content of the geographical, spatial entity of Europe, as present in the various vernacular translations and adaptations of the *Imago mundi*. The aim was to show that the boundaries, if they can be described as such, were fluid, and that even the seemingly objectively quantifiable value of geographical territory, with its 'natural boundaries' of, for instance, bodies of water, was subject of transformation in the process of translation between languages and cultures.[13] Indeed, water provides a suitable metaphor for the transformation of geographical boundaries, and ironically it is the European boundaries on land that are often described as

Age, 35 (Louvain-la-neuve: Féderation Internationale des Instituts d'Études Médiévales, 2006), p. 560.

11 For an illuminating discussion of the interplay between the 'linear' and 'zonal' meanings of 'frontier', see Nora Berend, 'Preface' to *Medieval Frontiers: Concepts and Practices*, ed. by David Abulafia and Nora Berend (Abingdon, 2016), pp. x–xvi.

12 Michael Wintle, *The Image of Europe: Visualising Europe in Cartography and Iconography Throughout the Ages* (Cambridge: Cambridge University Press, 2009), p. 35.

13 For water as natural geographical boundary of Europe, see Wintle, *The Image of Europe*, pp. 35–36.

fluid.[14] By contrast, the Mediterranean, in the *Imago mundi* itself, is not so much a boundary as a fourth part of the world, constituting a fuzzy set in its own right, on a par with Asia, Europe, and Africa. This is not maintained in most of its rewritings, which move the various islands to different part of the texts, as part of the major type of change they make: the change to the order of description, the way of leading the reader through the material, direction of perusal of the textual map. Along with the addition or updating of place-names, these constitute the major techniques of rewriting used in these texts.

The *Imago mundi* tradition also demonstrates the interrelation between the perception of space and the traversal of space, as the geographical description inside the text travels within the physical object of the manuscript through the geographical space it describes.[15] The 'mastery of space', to use Wetzstein's term, inherent in this movement, was a product of the activity of the monastic orders, and it is significant that the provenance of many of the manuscripts of the original Latin text of the *Imago mundi* points to the use of monastic networks in the transmission of this text.[16] However, it is also clear from the discussion above that the *Imago mundi* texts were not transmitted exclusively, and not even primarily, through ecclesiastical channels, with many of the rewritings heavily indebted to princely patronage. The spread of the *Imago mundi* tradition in Europe also echoes the pan-European network of intellectuals and their patterns of mobility and patronage in the period.[17]

This multilingual tradition also shows the applicability to a much broader period, of the characterisation given by Marie-Odille Garrigues to

14 See, for instance, discusssion in Wintle, *The Image of Europe*, pp. 37–38, and references therein.

15 Cf. Thomas Wetzstein's discussion of the connection between 'remote communication' and mobility; 'New Masters of Space: The Creation of Communication Networks in the West (Eleventh–Twelfth Centuries)', in *Space in the Medieval West*, ed. by Cohen and Madeline, pp. 115–32 (p. 119).

16 For more on medieval monastic networks and transmission of text, see, for instance, T. Falmagne, D. Stutzmann, A.-M. Turcan-Verkerk, and P. Gandil, eds, *Les cisterciens et la transmission des textes (XIII*e*–XVIII*e *siècles)* (Turnhout: Brepols, 2018); Wetzstein, 'New Masters', pp. 126–28 on the Cistercians; and most recently for Francistans in the Norse context, for instance, Johnny Grandjean Gøgsig Jakobsen, 'Friars of science: Dominican transmission and usage of scientific knowledge in medieval Scandinavia' in *Medieval Science in the North: Travelling Wisdom, 1000–1500*, ed. by Christian Etheridge and Michele Campopiano (Turnhout: Brepols, 2021), pp. 141–58.

17 See, for instance, William Chester Jordan, '"Europe" in the Middle Ages', in *The Idea of Europe: From Antiquity to the European Union*, ed. by Anthony Pagden (Cambridge: Cambridge University Press, 2002), pp. 72–90 (pp. 73–74).

Europe of the twelfth century: 'Europe sans frontière [...] où les hommes se déplacent et se rencontrent aussi facilement et plus profondément que de nos jours' ('Europe without borders [...] where people moved and met as easily and more profoundly than in our days').[18] Brunetto Latini provides an illustrative example of this. His *Trésor* is linked on the one hand to the *Imago mundi* itself, on the other to the *Image du monde*, and belongs to the same intellectual context of the francophone world. It also may have served as a model for the *Dittamondo*, for example. Brunetto travelled from Italy to Spain, and his connection to the court of Alfonso X brings his work closer (though not perhaps into direct contact) with the Spanish branch of the *Imago mundi* tradition, represented by the *Semeiança del mundo*, the Castilian translation of the *Imago mundi* pre-dating the grand Alfonsine programme of translation.[19] I have argued that the Italian map of the *Trésor* compared to the Latin source, with the focus on ecclesiastical rather than secular territories, is a reflection of the political circumstances of the text's production. The incorporation of a translation from *Imago mundi* into a different encyclopedia by another Italian notary, Vivaldo Belcalzèr, demonstrates that Latini's was not an isolated case, and the brief reflection on Rudolf of Ems's *Weltchronik*, which shows a similar approach, suggests that the refocusing and rewriting of *Imago mundi* for a new audience in this manner was not unique to Italy. This is supported by the parallels drawn earlier between the ecclesiastical focus of Brunetto Latini and Gervase of Tilbury's adaptations of geographical descriptions.[20] Like Latini's, both Rudolf's and Gervase's texts can be read in the context of the politics of the Holy Roman Empire. Since this is a study of Europe in *Imago mundi* texts, and not of Empire, these themes could only be briefly introduced here as topics for future research and must await their scholar(s).

The overlapping maps we have seen in our analysis of the Roman imperial provincial terminology used by Honorius and its overlay with ecclesiastical structures in the rewritings of Gervase and Brunetto Latini provide a model very similar to the model of historico-geographical reading proposed by Jostkleigrewe as a solution for the seeming confusion and overlap of areas in the *Image du monde*. The notion of overlapping maps, coupled with the

18 Marie-Odile Garrigues, 'Honorius Augustodunensis et l'Italie', *Mélanges de l'École Française de Rome. Moyen Âge, Temps Modernes*, 84 (1952), 511–30 (p. 512). The translation is mine.

19 It is worth noting, however, that the conventional date of this text is based on the arguments in the edition of 1959 and tends to be reproduced without question in newer discussions.

20 See above, pp. 101–02.

historico-geographical model (which we can also discuss in terms of the spatio-temporal model of modern physics and Bakhtin's *chronotope*) is also useful in thinking about the extended family of *Imago mundi* texts as a whole, since the vernacular rewritings were produced in areas where the Latin text was available and also continued to be copied alongside its subsequent adaptations and translations.[21]

In order to understand this phenomenon, we should take into account the fact that the transmission of the information carried in the *Imago mundi* into other cultural environments by means of rewritings does not constitute a process of replacement but rather of progressive growth. The geographical entity as it was imagined in this tradition underwent a number of transformations most of which, as we have seen, were dictated by the interests of the new audiences.[22] As Rita Copeland observes in relation to encyclopedias, the 'passage of the genre across languages can suggest how intellectual life too has a literary history, a history of form that adapts to new linguistic and cultural environments and that, with each new adaptation, signals changes in epistemological outlook'.[23] The dual effect of acceptance of the *Imago mundi* text as authoritative, and incorporation of minor alterations and refocusing of its geographical descriptions, echo the characteristically medieval tension identified by William Chester Jordan, which he describes as 'the tension between cosmopolitanism [...] and intense localism'.[24]

The refocusing and retargeting of descriptions of Europe, its shifting borders in the texts of the *Imago mundi* tradition examined in the present study, echo recent observations regarding the polyphony of identities within Europe from the twelfth century to the fifteenth. This illustrates Copeland's point that 'tracing certain textual forms from their Latin origins to their many vernacular ramifications, can offer us a way of breaking down [...] vast generalities bout the history of ideas'.[25] The local versions' refocusing of the descriptions of Europe, through the addition, for instance, of details relating to their own place of provenance to it, as did Caxton and Maerlant, amongst others, echoes Claire Weeda's argument regarding competition among the

21 For the *chronotope*, see above, pp. 24–28.

22 A connection is drawn between 'imaginations' and 'traditions' by Keith Lilley in his 'Introduction' to *Mapping Medieval Geographies: Geographical Encounters in the Latin West and Beyond, 300–1600*, ed. by Lilley (Cambridge: Cambridge University Press, 2013), pp. 1–20 (p. 2).

23 Rita Copeland, 'Language Frontiers, Literary Form, and the Encyclopedia', in *Frontiers in the Middle Ages*, ed. by Merisalo and Pahta, pp. 507–24 (p. 524).

24 Jordan, '"Europe" in the Middle Ages', p. 73.

25 Copeland, 'Language Frontiers', p. 523.

various elites.[26] In this family of texts, such competition is expressed by shifting the focus points of the Europe description. In this respect, though at first glance purely geographical, the definition of Europe in the *Imago mundi* texts is conncted to the much-debated issues of (territory-based) identity, including both the notion of Europeanness and national/ethnic identities. The precise ways in which one set of concepts may have influenced another or had a common source elsewhere would be a subject for a separate study.[27]

The thirteenth century, when many of the *Imago mundi*–based texts examined here were produced, has long been considered a turning point in medieval European political thought, particularly in thinking about territory and sovereignty.[28] It may well be worth examining the connections between this development and the increased production of vernacular rewritings of the *Imago mundi* tradition for princely patrons in the period and later. Such an endeavour may benefit from further integrating several important corpora that had to be excluded from the present study due to considerations of space. For instance, the use of the geographical frameworks discussed here in chronicles and historical texts produced throughout Europe in the medieval period may yield further data on the interrelation of space and time in the medieval worldview, and thus on the medieval *chronotope* of Europe, particularly in the context of discussion of 'national' histories.[29]

26 Claire Weeda, 'Images of Ethnicity in Later Medieval Europe' (unpublished PhD dissertation, Universiteit van Amsterdam, 2012), p. 11.

27 For a discussion of the interrelation between geographic areas and identity, see Weeda, 'Images of Ethnicity', pp. 26–27, 66–68.

28 Summarised by Chris Jones, 'Understanding Political Conceptions in the Later Middle Ages: The French Imperial Candidatures and the Idea of the Nation-State', *Viator*, 42 (2011), 83–114 (pp. 83–86), who proposes a re-evalutation. The perceived tension between universalism vs the idea of the sovereign state, particularly in Jones's nuanced reading, could be usefully taken in context of the geographical descriptions in the rewritings of the *Imago mundi* which I examine here.

29 E.g. in the *General Estoria*; see above, pp. 123–33. It may also be worthwhile to consider Europe in the medieval geographical tradition as no more and no less than one item in a tetrad of the fuzzy sets. As such, we might also keep in mind that approaches to understanding the medieval development of the geographical idea of Europe discussed in this book might also help develop an approach that could be fruitfully applied to the study of the other parts of the world. Both Asia and Africa have been treated in previous scholarship as concepts as problematic as that of Europe. We might take the history of the concepts of the geographical unit 'Africa' as an example. Francesc Relaño has recently argued that 'the idea of Africa' as a separate unit did not exist until the Renaissance, as in both Classical Antiquity and the Middle Ages, 'the southern shores of the Mediterranean were easily integrated into the classical *ecumene*', which he sees as a spatial continuum 'not fragmented into continents'; Relaño, *The Shaping of Africa: Cosmographic Discourse and Cartographic Science in Late Medieval and Early Modern Europe* (Aldershot: Ashgate, 2002), p. 1.

Investigations of these processes, and the role of traditional (antiquated) geographical information derived from the *Imago mundi* therein, would have to take into account the continued existence of the tradition I have introduced in this book in the early modern period, and the transition from manuscript to incunabula.

Looking Ahead

The decision to limit the narrative of this book to manuscript texts and take it only as far as the end of the fifteenth century is dictated more by convention and convenience than a belief in any great watershed at that point, either in terms of geographical writings or in terms of textual transmission.[30] However, the importance of the end of the fifteenth century in introducing changes, albeit gradual, in the way the *Imago mundi* and its worldview continued to influence minds, requires at least a short introduction. In that sense, the end of this book is the beginning of a different story.

One of the greatest changes to the European understanding of world geography came in the fifteenth century with the discovery of the Americas, but it came gradually. Even here one can argue that the process of development of geographical thought was one of evolution rather than of revolution. It is something of a commonplace that Columbus's worldview had been informed by clasical and medieval geography.[31] One of these texts which Columbus is known to have read, for we have an annotated copy surviving with notes in his own hand, is the *Ymago mundi* of Cardinal Pierre d'Ailly.[32]

30 In terms of the knowledge organisation, as Marlcom Parkes points out, there is more difference between early and late medieval books than between those and modern printed books, particularly in use of scholarly apparatus tables of contents, separation of the text into sections, use of notes; Parkes, 'The Influence of the Concepts of *Ordinatio* and *Compilatio* on the Development of the Book', in *Medieval Learning and Literature. Essays Presented to Richard William Hunt*, ed. by J. J. G. Alexander and M. T. Gibson (Oxford: Clarendon Press, 1976), pp. 115–41 (p. 135).

31 See Valerie I. J. Flint, *The Imaginative Landscape of Christopher Columbus* (Princeton, NJ: Princeton University Press, 1992); and Pauline Moffitt Watts, 'Prophecy and Discovery: On the Spiritual Origins of Christopher Columbus's "Enterprise of the Indies"', *The American Historical Review*, 90 (1985), 73–102.

32 The annotations are in Buron's edition; Edmond Buron, ed. and trans., *Ymago mundi de Pierre d'Ailly*, 3 vols (Paris: Maisonneuve Frères, 1930); Flint, *The Imaginative Landscape*, pp. 45–47; Watts, 'Prophecy'. Columbus refers to Pierre d'Ailly by name in the account describing his third voyage; *Select Letters of Christopher Columbus with Other Original Documents, Relating to His Four Voyages to the New World*, ed. by Richard Henry Major (Cambridge: Cambridge University Press, 1847, digitally reprinted in 2010), p. 140.

The importance of the geographical discoveries in the early modern period for the development of the concept of Europe is a subject for a different volume. Yet it must be noted here that, as Baumgärtner remarks, Europe became a more 'coherent space' with boundaries that came to need ever clearer demarcation as a response to the expansion of the outside world toward the end of the Middle Ages.[33] Nevertheless, representations of geographical space did not undergo an immediate revolution following the translation of Ptolemy's *Geographia* and Columbus's voyages.[34] It is for this reason that the term 'Renaissance' has been largely avoided in the present discussion. As has been judiciously observed recently by Patrick Gautier Dalché, the common representation of the Middle Ages opposed to 'modernity' is misleading as far as geographical representations of the world are concerned, and the rediscovery of Ptolemy did not mean the disappearance of other (e.g. medieval, T-O) modes of cartographic representation.[35] *Imago mundi* manuscripts such as Staats- und Stadtbibliothek Augsburg, 2 Cod 108 (s. xv, Eichstatt?) and various vernacular rewritings continued to be produced well into the period, alongside the creation of new rewritings and new adaptations (Figure 10). Some of the fifteenth-century adaptations, those by Juan de Mena and William Caxton in particular, were mentioned in this book. Caxton's is the only printed book discussed in *Transforming Europe*. I have made an exception in discussing this text because it is the only Middle English adaptation, and because it is closely connected to the *Image du monde*. The inclusion of these texts in this book is partly a tribute

33 Ingrid Baumgärtner, 'Europa in der Kartographie des Mittelalters: Repräsentationenen – Grenzen – Paradigmen', in *Europa im Weltbild des Mittelalters. Kartographische Konzepte*, ed. by Ingrid Baumgärtner and Hartmut Kugler (Berlin: Akademie Verlag, 2008), pp. 9–28 (p. 16).

34 See especially the discussion in Patrick Gautier Dalché, 'The Reception of Ptolemy's *Geography* (End of the Fourteenth to Beginning of the Sixteenth Century' in *The History of Cartography III: Cartography in the European Renaissance, Part I*, ed. by David Woodward (Chicago: University of Chicago Press, 2007), pp. 285–364, and the more extentive discussion in Dalché, *La géographie de Ptolémée en Occident (IVe–XVIe siècle)* (Turnhout: Brepols, 2009). The argument for the importance of the shifts in the representation on Europe, at the close of the Middle Ages, is put forward in Michael Wintle, 'Renaissance Maps and the Construction of the Idea of Europe', *Journal of Historical Geography*, 25 (1999), 137–65.

35 Patrick Gautier Dalché, '*Hic mappa mundi considerata est*: lecture de la mappemonde au Moyen Age', in *Itinerari del testo per Stefano Pittaluga*, ed. by C. Cocco et al. Publicazzioni del D.AR.FI.CL.ET. 'Francesco Della Corte', Third Series, 254 (Genova: Dipartimento di Antichità, Filosofia e Storia, 2018), pp. 495–516 (p. 506); see also Dalché, 'The Reception of Ptolemy's *Geography*', pp. 285–86. See also Natalia I. Petrovskaia, '地図を読む：ヨーロッパの世界地図、中世からルネサンスにかけて' (= 'Reading Maps: European World Maps, from the Middle Ages to the Renaissance'), 書物学 *Bibliology*, 13 (2018), 1–6.

to the continuity of the geographical tradition across artificially drawn boundaries of historical periodisation.

Incunabulae containing the *Imago mundi* and its adaptations constitute a significant corpus in themselves, and are yet to be studied. This tradition presents its own fascinating mysteries. Versions of Honorius's text were printed, and new translations produced as late as the eighteenth century. The editions are too numerous to produce a full catalogue here, but examples include Anton Koberger's version printed in Nuremberg in the fifteenth century, a later edition printed as part of *De Honorii Augustudunensis Presbyteri Libri Septem* in Basel, in 1544; one by Bernardus Albinus (1583) entitled *Mundi synopsis*, and many more up to the 1677 edition as part of the *Maxima Bibliotheca veterum Patrum* XX.[36] Several features of the earlier textual tradition, such as the misattributions to various authors continued well after the medieval period and made it into print. The misattribution to St Anselm of Canterbury is a case in point. *Imago mundi* forms part of Petrus Danhauser's edition of Anselm's *Opera* (published in Nuremberg by Caspar Hochfeder, 1491), as well as the edition of Anselm's works (published in Basel by Johann Amerbach, sometime before 1497).[37] This appears to be part of a wide pattern of misattribution, and the association of Mena's source, even when identified by title as the *Imago mundi*, with Anselm appears to belong to the same phenomenon.[38] This issue of misattribution deserves its own study.

Mena's *Laberinto* is in its turn one of many adaptations of the Latin text that went through multiple early modern print editions.[39] The *Image*

36 Edition ISTC number ih00323000; *British Library ISTC* https://data.cerl.org/istc/ih00323000 (accessed 11 April 2024). Annie Angremy, 'La *Mappemonde* de Pierre de Beauvais', *Romania*, 104 (1983), 316–50 (p. 319); France Autesserre, 'La vision du monde dans le *Laberinto de Fortuna* de Juan de Mena, réécriture de la représentation du monde dans le *De Imagine Mundi* d'Honorius Augustodunensis', *Cahiers d'études romanes*, 20, 'Traces d'autrui et retours sur soi' (2009), 127–70 (§§ 16–17).

37 Edition ISTC numbers ia00761000 and ia00759000; *British Library ISTC* https://data.cerl.org/istc/ia00761000 and https://data.cerl.org/istc/ia00759000 (accessed 11 April 2024).

38 The attribution of the *Imago mundi* to Anselm appears in the 1499 edition of Hernán Núñez and is maintained in subsequent scholarship; see discussion in Autesserre, 'La vision du monde', §§ 12–13.

39 Florence Street, 'La vida de Juan de Mena', *Bulletin Hispanique*, 55 (1953), 149–73 (p. 149). For in-depth discussion of the editions of the *Laberinto*, see Maxim P. A. M. Kerkhof and Rob Le Pair, 'El *Laberinto de Fortuna* de Juan de Mena: las ediciones en relación con la tradición manuscrita', in *Homenaje al Profesor Antonio Vilanova*, I, ed. by Marta Cristina Carbonell and Adolfo Sotelo Vázquez (Barcelona: Universidad de Barcelona, 1989), pp. 321–39; Maxim P. A. M. Kerkhof, 'Sobre las ediciones del *Laberinto de Fortuna* publicadas de 1481 a 1943, y la tradición manuscrita', in *Forum Litterarum: Miscelanea de estudos literários, linguisticos e históricos oferecida a J. J. van*

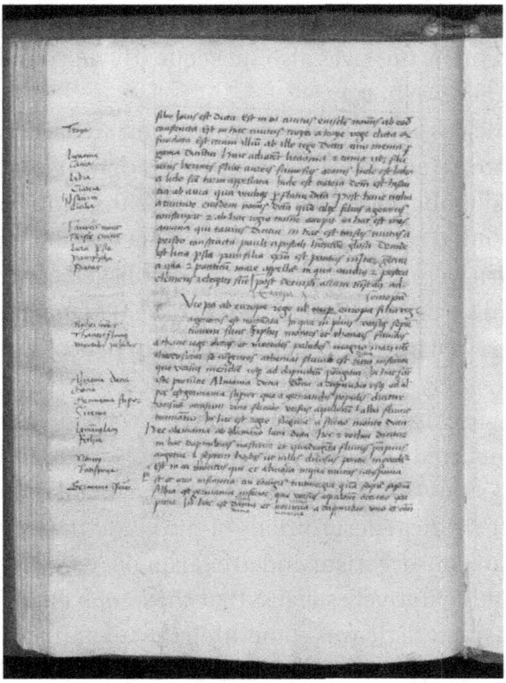

Figure 10: Beginning of the Europe section of the *Imago mundi* in Augsburg, Staats- und Stadtbibliothek Augsburg 2 Cod 108, f. 65v. Reproduced with kind permission of the Staats- und Stadtbibliothek Augsburg.

du monde, unsurprisingly given the number of surviving manuscripts, is another example.[40] Late-fifteenth-century editions include two by Antoine Caillaut (*c.* 1485–91 and *c.* 1491–92), Jean Trepperel (*c.* 1493–98).[41] Subsequent editions include *Le livre de clergie nommé l'Ymage du monde* (Paris, 1501) and J. Vivian's *Le mirouer du monde* (Geneva, 1517).[42] It is also the *Image du monde* that formed the basis of both the Hebrew and the Middle English

den Besselaar, ed. by J. A. H. Bots and Maxim P. A. M. Kerkhof (Amsterdam and Maarssen: APA and Holland University Press, 1984), pp. 26–82.

40 Neil Kenny mentions the Renaissance transmission of the *Image du monde*, and refers to its 1495, 1517, and 1520 editions (but describes it as an encyclopedia based on Vincent de Beauvais, not Honorius); Kenny, *The Palace of Secrets: Béroalde de Verville and Renaissance Conceptions of Knowledge* (Oxford: Oxford University Press, 1991), p. 18.

41 For more on these and references to four more sixteenth-century editions, see Christine Silvi, 'Petite histoire d'une appropriation réussie: François Buffereau plagiaire de Gossuin de Metz', *Revue belge de Philologie et d'Histoire*, 94 (2016), 661–89, n. 3; and for discussion of eight editions of the prose version, see Silvi, 'La mise en recueil de l'*Image du monde* (rédaction en prose) dans les premiers imprimés: diffusion et réception d'une *livre de clergie* médiéval aux XVe et XVIe siècles', *Seizième siècle*, 13 (2017), 293–328.

42 These editions are referred to in Chet Van Duzer, and Sandra Sáenz-López Pérez, '*Tres filii Noe diviserunt orbem post diluvium*: The World Map in British Library Add. MS 37049', *Word & Image*, 26 (2009), 21–39 (p. 32, p. 39 n. 92).

printed texts, which I have mentioned in this book. It is worth noting at this point that Caxton's *Mirror* in its turn was also subsequently reprinted with additions by Laurence Andrewe in 1527.[43]

Importantly in light of our discussion of the text's purposeful archaisim, this continuation of the printed tradition into the early modern period belies any notion of a post-Columbus watershed. Honorius's text and its rewritings continue being printed and being read, with new ones being produced. Some of these in turn were also subsequently republished many times. An example is Delacrut's צל העולם (*Ẓel ha-'Olam / Ṣel ha-'olam*), republished at least eleven times, well into the nineteenth century, mostly in Eastern Europe.[44]

The circulation of these materials in print represents a new chapter in the history of the *Imago mundi* tradition with its own complexities and must await its own dedicated study. It must be noted that the circulation of *Imago mundi*–derived geographical descriptions of the world in print illustrates the phenomenon of conservatism and archaism observed by scholars of incunabula. I would tentatively suggest that the *Imago mundi* did well here because this characteristic was something the text already had to start with. It fitted perfectly in the world of the antiquarian and conservative early printed book. The wide dissemination of this tradition in print supports the observation that it was more often material derived from ancient authorities and popular (old) traditions that would find its way to print, rather than innovative and revolutionary scientific advances.[45] Thus, the apparent break modern historiography places between medieval

43 *The Myrrour: [and] Dyscrypcyon of the Worlde* (London: Laurence Andrewe, 1527), replace link with: https://www.proquest.com/docview/2240950435?sourcetype=Books (acessed 27 November 2024). A. Neubauer, 'Les traductions hébraïques de l'*Image du Monde*', *Romania*, 5 (1876), 129–39 (p. 139), notes the existence of this edition, but does not identify it; see item 413 in C. E. Sayle, *Early English Printed Books in the University Library Cambridge (1475 to 1640)* (Cambridge: Cambridge University Press, 1900), I, pp. 85–86, and William E. Engel, Rory Loughnane and Grant Williams, eds, *The Memory Arts in Renaissance England: A Critical Anthology* (Cambridge: Cambridge University Press, 2016), pp. 41–42.

44 Abraham Melamed, 'The Hebrew Encyclopedias of the Renaissance', in Harvey, ed., *The Medieval Hebrew Encyclopedias of Science and Philosophy: Proceedings of the Bar-Ilan University Conference*, ed. by Steven Harvey, Amsterdam Studies in Jewish Thought, 7 (Dordrecht: Springer Science + Business Media, 2000), pp. 441–64; Robert Singerman, *Jewish Translation History: A Bibliography of Bibliographies and Studies* (Amsterdam: John Benjamins, 2002), p. 96 under item 679; and Gustavo Sacerdote, 'The Ninth Mehabbereth of Emanuele da Roma and the *Tresor* of Piere de Corbiac', *Jewish Quarterly Review*, 7 (1895), 711–28 (p. 711 n. 2).

45 Linda Ehrsam Voigts, 'Scientific and Medical Books', in *Book Production and Publishing in Britain 1375–1475*, ed. by Jeremy Griffiths and Derek Pearsall (Cambridge: Cambridge University Press, 1989), pp. 345–402, esp. p. 351, cited in Matthew Boyd Goldie, *Scribes of Space: Place in*

and early modern, the European encounter with the Americas, and the invention of print did not immediately affect the influence or cultural importance of the worldview represented by the *Imago mundi* tradition or its pattern of organisation.

The present study stops, with only a minor foray into the world of print, at the gates of the sixteenth century. In such a division lies the danger of reinforcing the view that manuscript culture was more fluid than that of the printed word, but some of the examples touched on in this book (such as Caxton's text) show that this view also needs to be qualified.[46] The increased mobility of the Renaissance, including the mobility of printers as well as scholars and artists, provides a context for the circulation of ideas and the dissemination of geographical and encyclopedic knowledge represented by the *Imago mundi* tradition.[47]

The continued evolution of the text even in printed form, evidenced by additions, alterations, corrections, and translations made after the Middle Ages, attest to a more flexible relation between text reception and text production than is usually represented in the dynamic of redactor–reader relations. Mary Franklin-Brown's 'more protean knowing "subject" that is opened up by encyclopedic texts', whom she characterises as 'neither a traditional author nor a passive reader' and who is 'actively engaged in the constitution of the text', is a useful concept for expressing this phenomenon.[48] Through the various adaptations, *Imago mundi* had an enormous influence. It is possible, for instance, that the etymology deriving the name of Europe from King Europs in Louis de Langle's *Tractatus de figura mundi* (1456) came indirectly from Honorius.[49]

Middle English Literature and Late Medieval Science (Ithaca, NY and London: Cornell University Press, 2019), p. 17, p. 226 n. 39.

46 Cf. Liedeke Plate and Els Rose in 'Rewriting, A Literary Concept for the Study of Cultural Memory: Towards a Transhistorical Approach to Cultural Remembrance', *Neophilologus*, 97 (2013), 611–25; Stephen D. Nichols, 'Introduction: Philology in a Manuscript Culture', *Speculum*, 65 (1990), 1–10 (p. 3).

47 For an overview of the circulation of knowledge in the sixteenth century, see Peter Burke, 'The Circulation of Knowledge', in *The Renaissance World*, ed. by J. J. Martin (New York: Routledge, 2007), pp. 191–207. The correlation between the movement of printers and the dissemination of *Imago mundi*–type texts in Europe would be a particularly interesting topic to explore further.

48 M. Franklin-Brown, *Reading the World: Encyclopedic Writing in the Scholastic Age* (Chicago: Chicago University Press, 2012), p. 26.

49 Among de Langle's sources Colette Beaune notes Latini's *Trésor* (which uses the *Imago mundi* but does not contain the etymology), Vincent de Beauvais and Pierre d'Ailly, but not the *Imago mundi* itself; Beaune, 'La notion d'Europe dans les livres d'astrologie du XVe siècle', in *La conscience européenne au XVe et au XVIe siècle*, Collection de l'Ecole Normale Supérieure de Jeunes Filles, 22 (Paris : ENSJF, 1982), pp. 1–7 (pp. 3–4).

This book may be more accurately entitled 'Transforming Europes', as its chief finding is the polyphony of voices within what might have been expected to be a uniform and static encyclopedic tradition. This multitude of voices provides a multiphasic representation of a geographical territory with shifting borders and shifting focus points. Combining these into a single term produces the textual equivalent of an out-of-phase image, which might be compared to stereoscopic layers viewed without 3D glasses. Coupled with the notion of fuzzy sets this results in a delightfully unfocused picture. Each rewriting of *Imago mundi* produced its own, slightly different Europe, and thus the broader *Imago mundi* tradition encompassed and transmitted to new readers a multitude of Europes.

Appendices

Appendix 1: List of *Imago Mundi* chapters

The list below presents the chapters of *Imago mundi*, Book I, with titles. The titles are taken from Flint, ed., *Imago mundi*, which takes chapter divisions and headings from Munich, Bayerische Staatsbibliothek clm. 536 (Prül, s. XII).[1] Although the references to chapter numbers throughout this book are to Flint's numbering, since studies that pre-date her edition relied on the chapter numbering in Migne's PL edition, this is also provided (in square brackets).

Epistola cuiusdam ad Honorium solitarium (Letter of Christianus)
Prologus solitarii Christiano (Letter of Honorius)
1[1] *De forma mundi*
2[2] *De creatione mundi*
3[3] *De elementis*
4[4] *De .vii. nominibus terre*
5[5] *De forma terre*
6[6] *De .v. zonis*
7[7] *De .iii. partibus*
8[8, 9] *De paradyso*
9[10] *De .iiii. fluminibus*[2]
10[11] *De India*
11[12] *De Monstris*[3]
12[13] *De bestiis*[4]
13[14] *De Parthia*
14[15] *De Mesopotamia*
15[16] *De Syria*[5]
16[17] *De Palestina*
17[18] *De Egypto*
18[19] *De regionibus orientis*[6]
19[20] *De Asia Minor*[7]
20[21] *De regionibus Asie*[8]
21[22] *De Europa*
22[23] *De Scithia*
23[24] *De Germania Superior*
24[25, 26] *De Germania Inferiore*

1 Honorius Augustodunensis, *Imago mundi*, ed. by Valerie I. J. Flint, *Archives d'histoire doctrinale et littéraire du Moyen Âge*, 49 (1982), p. 45.

2 *'De De .iiii. fluminibus'* in Flint's edition; presumably a typographical error.

3 A chapter concerning humanoid beings, including Sciapods, Cyclopes, Pygmies.

4 Concerning animals, largely fantastical, corresponding in their description to the Bestiary tradition, ultimately derived from Solinus (e.g. *Ceucocrota*, the Yale, the *Monoceros*) but also including elephants and turtles. For a discussion, see Natalia I. Petrovskaia, 'L'image du monde animalier', in *Mondes animaliers au Moyen Age et à la Renaissance – Tierische Welten im Mittelalter und in der Renaissance – Actes du Colloque international des 8, 9, 10 et 11 mars 2016 à la Maison de la Culture d'Amiens*, ed. by Danielle Buschinger, Florent Gabaude, Marie-Geneviève Grossel, Jürgen Kühnel, and Mathieu Olivier (Amiens: Presses du 'Centre d'Etudes Médiévales de Picardie', 2016), pp. 313–26.

5 This chapter includes a discussion of Jerusalem.

6 Includes description of the Caucasus, the Amazons, and the regions of Bactria and Hircania, Albania, Capadocia, and Armenia (with mention of Mount Ararat).

7 With mention of Bythinia, Nicomedia, among other regions.

8 From Bythinia and Phrygia, to Pamphylia and the Hellespont. Reference to the place of exile of Ovid and Pope Clement I.

25[27] *De Grecia*
26[28] *De Italia*
27[29] *Gallia*
28[30] *Hispania*
29[31] *Britannia*[9]
30[32] *Affrica*
31[32] *Kartago*
32[33] *Ethiopia*
33[34] *De insulis*
34[35] *Sicilia*
35[36] *Sardinia*
36[37] *De inferno*
37[37] *De nominibus inferni*
38[38] *De aqua*
39[39] *De oceano*
40[40] *De estu*
41[41] *De voragine*
42[42] *De terre motu*
43[43] *De hiatu*
44[43] *De Sicilia*
45[43] *De Ethna*
46[43] *De Scilla*
47[44] *De frigore*
48[45] *De dulcibus aquis*
49[45] *De amaris*
50[46] *De Mari Rubro*
51[46] *De mari*
52[47] *De gemina natura aquarum*
53[48] *De aqua calida*
54[49] *De mortifera aqua*
55[50] *De Mortuo Mari*
56[51] *De animalibus aquarum*
57[52] *De signis*
58[53] *De aere*
59[54] *De vento*
60[55] *De cardinalibus*
61[56] *De nubibus*
62[56, 57] *De tonituro et fulgure*
63[58] *De iri*
64[59] *De pluvia*
65[60] *De grandine*
66[61] *De nive*
67[62] *De rore et pluvia*
68[63] *De nebula*
69[64] *De fumo*
70[65] *De igniculo*
71[66] *De pestilentia*
72[67] *De igne*
73[68] *De planetis*
74[69] *De luna*
75[70] *De Mercurio*
76[71] *De Venere*
77[72] *De Sole*
78[73] *De signis*
79[74] *De Marte*
80[75] *De Iove*
81[76] *De Saturno*
82[77] *De absidibus*
83[78] *De colore*
84[79] *De via illorum*
85[80] *De sono illorum*
86[81] *Celestis Musica*
87[82] *De homine*
88[83] *De mensura*
89[84] *De celo*
90[84] *De ianuis*
91[85] *De clymatibus*
92[86] *De plagis*
93[87] *De firmamento*
94[88] *De axe*
95[89] *De stellis*
96[90] *De syderibus*
97[91] *De zodiaco*
98[92] *De signis. Aries*
99[93] *Taurus*
100[94] *Gemini*
101[95] *Cancer*
102[96] *Leo*
103[97] *Virgo*
104[98] *Libra*
105[99] *Scorpius*
106[100] *Sagittarius*
107[101] *Capricornus*
108[102] *Aquarius*
109[103] *Pisces*
110[104] *Hiades*
111[105] *Pliades*
112[106] *Arcton*
113[107] *Artophilax*
114[108] *Arcturus*
115[109] *Phiton*
116[110] *Corona*
117[111] *Hercules*
118[112] *Lira*
119[113] *Cignus*
120[114] *Cepheus et Casiepea*
121[115] *Perseus et Andromeda*
122[116] *Deltoton*
123[117] *Serpentarius*
124[118] *Pegasus*
125[119] *Delfin*
126[120] *Aquila*
127[121] *Sagitta*
128[122] *Ydra*
129[123] *Crater*
130[124] *Corvus*
131[125] *Orion*
132[126] *Anticanis*

9 Chapter includes the British Isles and Ireland.

133[127] *Canis*
134[128] *Lepus*
135[129] *Eridanus*
136[130] *Cetus*
137[131] *Centaurus*
138[132] *Ara*
139[133] *Argo*
140[134] *Pistrix*
141[135] *Canopus*
142[135] *Hemispherum*
143[136] *Lactea via*
144[137] *De cometa*
145[138] *Aqueum celum*
146[139] *Spiritale celum*
147[140] *Celum celi*

Appendix 2: Texts of the *Imago Mundi* Tradition

This list includes only texts which make demonstrable use of *Imago mundi* I, in the form of lengthy (sometimes translated) quotations (the whole book or at least multiple chapters). The list is not all-inclusive, and its intention is to provide insight into the extent of the multilingual *Imago mundi* tradition, as well as a reference point to anchor the discussion in this book. Translations, adaptations, and works quoting *in extenso* or heavily based on the *Imago mundi* discussed in this book are included. For the sake of consistency, only the century is given as the date here, since not all of the texts have been (or can be) precisely dated. The chronological order given is therefore also approximate.

Pierre de Beauvais, *Mappemonde* (French), s. XIII
Anon., *Semeiança del mundo* (Spanish), s. XIII
Gervase of Tilbury, *Otia imperialia* (Latin), s. XIII
Anon., *La Petite Philosophie* (Anglo-Norman), s. XIII
Anon., *Delw y Byd* (Welsh), two, possibly three separate translations, s. XIII
Gossouin/Gautier de Metz, *Image du monde* (French), two, possibly three versions, s. XIII
Jacob van Maerlant, *Alexanders geesten* (Book VII, ll. 837–1776) (Dutch), s. XIII
Anon., *Ymagine del mondo* (Italian), multiple versions, s. XIII
Brunetto Latini, *Trésor* (French), and *Tresoretto* (Italian), s. XIII
Anon., *Divisiuns del mund / Divisiones mundi* (Anglo-Norman), s. XIII?/XIV
Vivaldo Belcalzér, *Mapa del mond* (Italian), s. XIV
'Catalan Atlas', BnF, MS Esp. 30 (Catalan), s. XIV
Pierre d'Ailly, *Ymago mundo* (Latin), s. XV
Juan de Mena, *Laberinto de Fortuna* (Spanish), s. XV
Anon., *Das puch von der Forme und pildnuss der Welt* (German), s. XV^{ex}
William Caxon, *Mirror of the World* (English), s. XV
David ben Moses *Demuth ha'olam* (Hebrew), date unknown
Mattathias ben Solomon Delacrut (attr. to Hagin Deulacres), צל העולם *Ẓel ha-'Olam / Ṣel ha-'olam / Tsel ha-'olam* (Hebrew), s. XVI

Bibliography

Manuscripts

Augsburg, Staats- und Stadtbibliothek Augsburg, 2 Cod108 (Eichstatt?, s. XV)
Baltimore, MD, Walters Museum of Art, MS W.199 (Bruges, 1489)
Berlin, Staatsbibliothek Preussischer Kulturbesitz, Germ. Fol. 595 (Niederbayern, s. XIV/XV)
Cambridge, Corpus Christi College, MS 66 (Sawley, s. XII)
Copenhagen, Det Kongelige Bibliothek, Gml. kgl. Sml. 1812 4to
Copenhagen, University of Copenhagen, AM 655XX 4to (s. XIV)
Ghent, University Library, MS 92 (Saint-Omer, s. XII)
Lisbon, Biblioteca Nacional, MS 46 (s. XIII)
London, BL, MS Harley 4333 (s. xiii)
London, Lambeth Palace, MS 371 (Reading Abbey, s. XIII)
Madrid, Biblioteca Nacional, MS 3369 (s. XIV–XV)
Madrid, Biblioteca del Escorial, X.iii.4 (1467)
Modena, Biblioteca Estense, MS α.Q.5.1 (s. XIV)
Munich, Bayerische Staatsbibliothek, clm. 536 (Prül, s. XII)
Oxford, Bodleian Library, MS Bodley 283 (s. XV)
Oxford, Bodleian Library, MS Rawlinson B.484 (s. XII$^{ex.}$)
Oxford, Jesus College, MS 20 (s. XIV^{2})
Palma de Mallorca, Biblioteca de Fondación Bartolomé March, MS B80-B-17 (s. XV)
Paris, BnF, MS Esp. 30 (s. XIV)
Paris, BnF, MS Fr. 14961 (Limousin, s. XIII)
Reykjavik, Árni Magnússon Institute, AM 142 fol. (1690–1697)
Reykjavik, Árni Magnússon Institute, AM 566 c 4to (1705)
Reykjavik, Árni Magnússon Institute AM 685d 4to, f. 31r (s. XV)
Reykjavik, Árni Magnússon Institute AM 746 4to, ff. 2r–v (s. XIV)
Reykjavik, Árni Magnússon Institute, AM 764 4to (s. XIV)
Reykjavik, Árni Magnússon Institute, AM 238 fol. XIX, ff. 1r–v (s. XV)
Vienna, Österreichische Nationalbibliothek, MS 382 (Lambach, s. XII)

Primary Sources

Alfonso el Sabio, *General estoria: Part 1*, ed. by Antonio G. Solalinde (Madrid: Centro de Estudios Históricos, 1930)

Alfonso el Sabio, *General estoria: Part 2*, ed. by Antonio G. Solalinde, Lloyd A. Kasten, and Victor R. B. Oelschläger, 2 vols (Madrid: Consejo Superior de Investigaciones Científicas, 1957–61)

Alfonso X El Sabio, *General Estoria: VI Partes*, ed. by Pedro Sánchez-Prieto, 10 vols (Madrid: Biblioteca Castro, Fundación José Antonio de Castro, 2009)

Alfonso X, *Lapidario (según el manuscrito Escurialense H.I.15)*, ed. by Sagrario Rodríguez M. Montalvo (Madrid: Editorial Gredos, 1981)

Alfræði Íslenzk, ed. by K. Kålund and N. Beckman, 3 vols (København: Samfund til Udgivelse af gammel nordisk Litteratur, 1908–1918)

Andrewe, Laurence, *The Myrrour: [and] Dyscrypcyon of the Worlde* (London, 1527)

Amerbach, Johann, *Opuscula beati Anselmi archiepiscopi Cantuariensis* (Basel, *ante* 1497)

Apian, Peter, *Cosmographia, siue Descriptio universi orbis* (Antwerp: Jan Bellerus, 1584)

Ariosto, Ludovico, *Orlando Furioso*, trans. by Guido Waldman, Oxford World Classics (Oxford: Oxford University Press, 1974)

Ariosto, Ludovico, *Orlando Furioso: A Selection*, ed. by Pamela Waley (Manchester: Manchester University Press, 1975)

Barney, Stephen A., W. J. Lewis, J. A. Beach, and Oliver Berghof, trans., *The Etymologies of Isidore of Seville* (Cambridge: Cambridge University Press, 2006)

Bede, *On the Nature of Things and On Times*, trans. by Calvin B. Kendall and Faith Wallis, Translated Texts for Historians, 56 (Liverpool: Liverpool University Press, 2010)

Brunetto Latini, *Il tesoretto: The Little Treasure*, ed. and trans. by J. B. Holloway, Garland Library of Medieval Literature, 2 (New York: Garland, 1981)

Brunetto Latini, *The Book of Treasure* (*Li Livres dou Tresor*), trans. by Paul Barrette and Spurgeon Baldwin, Garland Library of Medieval Literature Series B, 90 (New York and London: Garland, 1993)

Brunetto Latini, *Li Livres dou Tresor*, ed. by Francis Carmody (Berkeley: University of California Press, 1948)

Brunetto Latini, *Tresor*, ed. by Pietro G. Beltrami, Paolo Squilacioti, Plinio Torri, and Sergio Vatteroni (Turin: Einaudi, 2007)

Bull, William E., and Harry F. Williams, eds, *Semeiança del Mundo: A Medieval Description of the World* (Berkeley and Los Angeles: University of California Press, 1959)

Buron, Edmond, ed. and trans., *Ymago mundi de Pierre d'Ailly*, 3 vols (Paris: Maisonneuve Frères, 1930)

Calendarium Historico-Christianum medii et novi aevi, ed. by A. J. Weidenbach (Regensburg, 1855)

Caxton, William, *Myrrour of the World* (Westminster: William Caxton, 1481; 2nd edn *c.* 1490)

Caxton, William, *Mirrour of the World*, ed. by Oliver H. Prior, Early English Texts Society e.s., 110 (London: Kegan Paul, Trench, Trübner, 1913)

Chiovaro, Francesco, ed., *L'Ymagine del mondo* (*Firenze, Bibl. naz. cod. palat. 703*), Quaderni partenopei, 1 (Naples: Loffredo, 1977)

Corpus diacrónico del español (CORDE), Real Academia Española, Banco de datos (online) http://www.rae.es

Dalché, Patrick Gautier, ed., *Carte marine et portulan au XIIe siècle: Le Liber de Existencia Riverierarum et Forma Maris Nostri Mediterranei* (*Pise, circa 1200*) (Rome: l'École Française de Rome, 1995)

Danhauser, Petrus, *Opera et tractatus beati Anselmi* (Nuremberg: Caspar Hochfeder, 1491)

Dati, *La sfera: Libri Quattro in ottava rima*, ed. by Enrico Narducci (Milan, 1865; reprinted Bologna, 1975)

Dunphy, R. Graeme, ed. and trans., *History as Literature: German World Chronicles of the Thirteenth Century in Verse. Excerpts from: Rudolf von Ems,* Weltchronik, *The* Christherre-Chronik, *Jans Enikel,* Weltchronik, Medieval German Texts in Bilingual Editions, 3 (Kalamazoo, MI: Medieval Institute Publications, Western Michigan University: 2003)

Fazio degli Uberti, *Il Dittamondo e le rime*, ed. by Giuseppe Corsi, 2 vols (Bari: Giuseppe Laterza & Figli, 1952)

Finnur Jónsson, ed., *Rímnasafn: Samling af de aldeste islandske rimer*, 2 vols (Copenhagen: Møller & Jørgensen, 1905–1922)

Finzi, Vittorio, 'Di un inedito volgarizzamento dell'*Imago mundi* di Onorio d'Autun', *Zeitschrift für Romanische Philologie*, 17 (1893), 490–543; 18 (1894), 1–73

Firchow, Evelyn Scherabon, ed. and trans., *The Old Norse 'Elucidarius': Original Text and English Translation*, Medieval Texts and Translations (Columbia, SC: Camden House, 1992)

Foster, Nicholas Ryan, 'The *Imago mundi* of Honorius Augustodunensis' (unpublished MA thesis, Portland State University Department of History, 2008), DOI: 10.15760/etd.5974 (accessed 11 November 2024)

Gervase of Tilbury, *Otia Imperialia: Recreation for an Emperor*, ed. and trans. by S. E. Banks and J. W. Binns (Oxford: Clarendon Press, 2002)

Gilbertus Porretanus, *In librum (Boetii) de duabus naturis et una persona Christi commentaria*, ed. by J.-P. Migne, PL, 64 (Paris, 1847), cols 1333–1412

Gossouin de Metz, *L'Image du monde, une encyclopédie du XIIIe siècle: edition critique et commentaire de la première version*, ed. by Chantal Connochie-Bourgne (unpublished PhD thesis, University of Paris, Sorbonne, Paris IV, 1999)

Honorii Augustodunensis Presbyteri Libri Septem (Basel: Andreas Cratander, 1544)

Honorius Augustodunensis, *Elucidarium*, ed. by Yves Lefèvre, *'L'Elucidarium' et les lucidaires : Contribution, par l'histoire d'un texte, à l'histoire des croyances religieuses en France au Moyen Âge*, Bibliothèque des Écoles françaises d'Athènes et de Rome, 180 (Paris: E. de Brocard, 1954), pp. 343–497

Honorius Augustodunensis, *Imago mundi*, ed. by Valerie I. J. Flint, *Archives d'histoire doctrinale et littéraire du Moyen Âge*, 49 (1982)

Honorius Augustodunensis, *De Luminaribus ecclesiae*, ed. by J.-P. Migne, PL, 172 (Paris, 1895), 232–234

Honorius Augustodunensis, *De solis affectibus*, ed. by J.-P. Migne, PL, 172 (Paris, 1895), cols 102–120

Honorius Augustodunensis, *Summa totius*, ed. by J.-P. Migne, PL, 172 (Paris, 1895), cols 187–196

Isocrates, *Isocrates*, 1, with translation by George Norlin, Loeb Classical Library, 209 (Cambridge MA and London: Harvard University Press, 1945)Jacob van Maerlant, *Alexanders geesten*, ed. by Johannes Franck (Groningen: J. B. Wolters, 1882)

Jensen, Helle, ed., *Eiríks saga víðförla* (Copenhagen: Reizel, 1983)

Jónas Kristjánsson, ed., *Dínus saga drambláta*, Riddarsōgur, 1 (Reykjavik: Háskóli Íslands, 1960)

Juan de Mena, *Laberinto de Fortuna*, ed. by John G. Cummins (Madrid: Cátedra, 1979)

Juan de Mena, *Laberinto de Fortuna*, ed. by Maxim P. A. M. Kerkhof, Nueva Biblioteca de Erudición y Crítica, 9 (Madrid: Castalia, 1995)

Kasten, Lloyd, John Nitti, and Jean Anderson, *Concordances and Texts of the Royal Scriptorium Manuscripts of Alfonso X, el Sabio* (Madison: Hispanic Seminary of Medieval Studies, 1978)

Lewis, Henry, and P. Diverres, ed., *Delwy Byd (Imago Mundi)* (Cardiff: Cardiff University Press, 1928)

Lindsay, W. M., ed., *Isidori Hispalensis Episcopi Etymologiarvm sive Originvm Libri XX*, 2 vols (Oxford: Oxford University Press, 1911)

Major, Richard Henry, ed., *Select Letters of Christopher Columbus with Other Original Documents, Relating to His Four Voyages to the New World* (Cambridge: Cambridge University Press, 1847)

Matteo Maria Boiardo, *L'inamoramento de Orlando*, ed. by Antonia Tissoni Benvenuti and Christna Montagnani, 2 vols (Milan and Naples: Ricciardi, 1999)

Maxima Bibliotheca Veterum Patrum, vol. XX (Scriptores 1120–1130) (Lyon, 1677)

Migne, Jean-Paul, ed., *Honorii Augustodunensi De imagine mundi libri tres*, PL172 (Paris, 1895)

Mundi Synopsis siue De Imagine Mundi Libri Tres (Spirae ciuitate veterum Nemetum: apud Bernardum Albinum, 1583)

Núñez, Hernán, ed., *Las .CCC. del famosissimo poeta Juan de Mena con glosa* (Seville, 1499)

Orosius, Paulus, *Pavli Orosii Historiarum Adversum Paganos libri VII*, ed. by Karl Zangemeister (Vienna: apud Geroldi Filium Bibliopolam Academiae, 1882)

Petrovskaia, Natalia I., ed., *Delwy Byd: A Medieval Welsh Encyclopedia*, MHRA Library of Medieval Welsh Literature (London: MHRA, 2020)
Pliny, *Natural History, Books 3–7*, with translation by H. Rackham, Loeb Classical Library, 352 (Cambridge MA and London: Harvard University Press, 1942)Poppe, E., and R. Reck, ed., *Selections from Ystorya Bown o Hamtwn* (Cardiff: University of Wales Press, 2009)
Primera Crónica General de España [= *Estoria de Espanna*], 2 vols., ed. by Ramón Menéndez Pidal (Madrid: Editorial Gredos – Cátedra Seminrio Menéndez Pidal, 1979)
Prior, O. H., ed., *Divisiones Mundi*, in *Cambridge Anglo-Norman Texts* (Cambridge: Cambridge University Press, 1924), pp. 33–62
Prior, O. H., ed., *L'Image du monde de maitre Gossuin: Rédaction en prose* (Lausanne and Paris, 1913)
Ranulf of Higden, *Polychronicon Ranulphi Higden Monachi Cestrensis; together with the English Translation of John Trevisa and of an Unknown Writer of the Fifteenth Century*, ed. by Churchill Babington and Joseph R. Lunby, Rolls Series, 9 vols (London: Her Majesty's Stationary Office, Rolls Commission, 1865–1886)
Solinus, *Collectanea rerum memorabilium*, ed. by Th. Mommsen (Berlin: Weidmann, 1895)
Trethewey, W. H., ed., *La Petite Philosophie, An Anglo-Norman Poem of the Thirteenth Century: Text with introduction, notes and glossary*, Anglo-Norman Texts, 1 (Oxford: Anglo-Norman Text Society, 1939)
William of Malmesbury, *De Gestis Regum Anglorum*, ed. by W. Stubbs, Rolls Series, 90, 2 vols (London: Her Majesty's Stationary Office, Rolls Commission, 1887–89)
Wilmans, D. R., ed., *Ex Honorii Augustodunensis Summa Totius et Imagine Mundi*, Monumenta Germaniae Historica: Scriptorum, 10 (Hannover, 1852)
Wittlin, Curt J., *Llibre del tresor: versió catalana de Guillem de Copons* (Barcelona: Barcino, 1980–89), 4 vols.

Secondary Sources

Abulafia, David, *The Great Sea: A Human History of the Mediterranean* (Oxford: Oxford University Press, 2011)
Abulafia, David, *Frederick II: A Medieval Emperor* (Oxford: Oxford University Press, 1988)
Abulafia, David, review of Allen James Fromherz, *The Near West: Medieval North Africa, Latin Europe and the Mediterranean in the Second Axial Age* by (2016), *Journal of Islamic Studies*, 29 (2018), 110–12
Abulafia, David, and Nora Berend, eds, *Medieval Frontiers: Concepts and Practices* (New York and Abingdon: Routledge, 2016; first ed. Aldershot: Ashgate, 2002)
Akbari, Suzanne Conklin, *Idols in the East: European Representations of Islam and the Orient, 1100–1450* (Ithaca, NY and London: Cornell University Press, 2009)
Almagià, Roberto, *Planisferi, carte nautiche e affini dal secolo XIV al XVII esistenti nella Biblioteca Apostolica Vaticana*, Monumenta Cartographica Vaticana, 1 (Vatican: Biblioteca Apostolica Vaticana, 1944)
Amory, Frederic, 'Things Greek and the Riddarasögur', *Speculum*, 59 (1984), 509–23
Angremy, Annie, 'La *Mappemonde* de Pierre de Beauvais', *Romania*, 104 (1983), 316–50
Armstrong, Adrian, and Sarah Kay, with participation of Rebecca Dixon, Miranda Griffin, Sylvia Huot, Francesca Nicholson, and Finn Sinclair, *Knowing Poetry: Verse in Medieval France from the 'Rose' to the 'Rhétoriques'* (Ithaca, NY and London: Cornell University Press, 2011)

Asbach, Olaf, *Europa – Vom Mythos zur 'Imagined Community'? Zur historischen Semantik 'Europas' von der Antike bus ins 17. Jahrhundert*, Europa und Moderne, 1 (Munich: Wehrhahn, 2011)

Ashhurst, David, 'Imagining Paradise', in *The Fantastic in Old Norse/Icelandic Literature: Sagas and the British Isles. Preprint papers of the 13th international Saga Conference, Durham and York, 6th–12th August, 2006*, ed. by John McKinnell, David Ashurst, and Donata Kick (Durham: Durham University Press, 2006), pp. 71–80

Atalay, Bülent, *Math and the Mona Lisa: The Art and Science of Leonardo da Vinci* (New York: HarperCollins and the Smithsonian Institution, 2006)

Autesserre, France, 'La vision du monde dans le *Laberinto de Fortuna* de Juan de Mena, réécriture de la représentation du monde dans le *De Imagine Mundi* d'Honorius Augustodunensis', *Cahiers d'études romanes*, 20 'Traces d'autrui et retours sur soi' (2009), 127–70

Bakhtin, Mikhail M., 'Forms of Time and of the Chronotope in the Novel: Notes toward a Historical Poetics', in *The Dialogic Imagination: Four Essays by M. M. Bakhtin*, ed. by Michael Holquist, trans. by Caryl Emerson and Michael Holquist, University of Texas Press Slavic Series, 1 (Austin: University of Texas Press, 1981), pp. 84–258

Bakirtzis, Nikolas, and Xenophon Monairos, 'Mastic Production in Medieval Chios: Economic Flows and Transitions in an Insular Setting', *Al-Masāq: Journal of the Medieval Mediterranean*, 31 (2019), 171–95.

Balard, Michel, 'The Greeks of Crimea under Genoese Rule in the XIVth and XVth Centuries', *Dumbarton Oaks Papers*, 49 (1995), 23–32

Bamfi, Florio, 'The Cosmographic Loggia of the Vatican Palace', *Imago mundi*, 9 (1952), 23–34

Barber, Malcolm, 'Western Attitudes to Frankish Greece in the Thirteenth Century', *Mediterranean Historical Review*, 4 (1989), 111–28

Bartlett, Robert, 'Heartland and Border: The Mental and Physical Geography of Medieval Europe', in *Power and Identity in the Middle Ages: Essays in Memory of Rees Davies*, ed. by Huw Pryce and John Watts (Oxford: Oxford University Press, 2007), pp. 23–36

Bartuschat, Johannes 'La forma allegorica del "Tesoretto" e il "Dittamondo" di Fazio degli Uberti', in *A scuola con Ser Brunetto: la ricezione di Brunetto Latini dal Medioevo al Rinascimento*, ed. I. Maffia Scariati (Florence: SIMSEL–Edizioni del Galluzzo, 2008), pp. 417–35

Bassnett, Susan, 'From Cultural Turn to Translational Turn: a Transnational Journey', in *Literature, Geography, Translation: Studies in World Writing*, ed. by Cecilia Alvstad, Stefan Helgesson, and David Watson (Newcastle upon Tyne: Cambridge Scholars Publishing, 2011), pp. 67–80

Baumgärtner, Ingrid, 'Europa in der Kartographie des Mittelalters. Repräsentationenen – Grenzen – Paradigmen', in *Europa im Weltbild des Mittelalters: Kartographische Konzepte*, ed. by Ingrid Baumgärtner and Hartmut Kugler (Berlin: Akademie Verlag, 2008), pp. 9–28

Baumgärtner, Ingrid, and Hartmut Kugler, eds, *Europa im Weltbild des Mittelalters: Kartographische Konzepte* (Berlin: Akademie Verlag, 2008)

Baxter, Jason, '"Videmus nunc per speculum": The Mysticism and Naturalism of the Twelfth-Century *imago mundi*', *Haskins Society Journal*, 28 (2017), 119–42

Beaune, Colette, 'La notion d'Europe dans les livres d'astrologie du XVe siècle', in *La conscience européenne au XVe et au XVIe siècle*, Collection de l'Ecole Normale Supérieure de Jeunes Filles, 22 (Paris: ENSJF, 1982), pp. 1–7

Becker, Peter Jörg, and Eef Overgaauw, eds, *Aderlass und Seelentorst: die Überlieferung deutscher Texte im Spiegel Berliner Handschriften und Inkunabeln* (Mainz am Rhein: Verlag Philipp von Zabern for Staatsbibliothek zu Berlin, 2003)

Belliato, Nadia, 'Per un commento al "Dittamondo": il Paradiso terrestre e la personificazione di Roma (I.xi)', *Quaderno di italianistica 2015*, Sezione di Italiano dell'Università di Losana (Pisa: ETS, 2015), pp. 33–48

Beltrami, Pietro G., 'Appunti su vicende del *Tresor*: composizione, letture, riscritture', in *L'enciclopedismo medievale*, ed. by Picone, pp. 311–28

Berend, Nora 'The Concept of Christendom: A Rhetoric of Integration of Disintegration?', in *Hybride Kulturen im mittelalterlichen Europa*, ed. by Michael Borgolte and Bernd Schneidmüller (Berlin: de Gruyter, 2010), pp. 51–62

Berend, Nora, 'Preface' to *Medieval Frontiers: Concepts and Practices*, ed. by Abulafia and Berend, pp. x–xvi

Berendrecht, Petra, *Proeven van bekwaamheid: Jacob van Maerlant en de omgang met zijn Latijnse bronnen* (Amsterdam: Prometheus, 1996)

Bergman, Merrie, *An Introduction to Many-Valued and Fuzzy Logic: Semantics, Algebras and Derivation Systems* (Cambridge: Cambridge University Press, 2008)

Besamusca, A. A. M., 'Courtly Literature in the Low Countries and Germany: Jacob van Maerlant and Rudolf von Ems', in *Dutch and Flemish literature as world literature*, ed. by T. D'haen, Literatures as World Literature (New York: Bloomsbury Academic, 2019), pp. 44–53

Beyer de Ryke, Benoît, 'Le miroir du monde: un parcours dans l'encyclopédisme médiéval', *Revue belge de philologie et d'histoire*, 81 (2003), 1243–75

Biglieri, Aníbal A., *Las ideas geográficas y la imagen del mundo en la literatura española medieval*, Medievalia Hispanica, 17 (Madrid and Frankfurt am Main: Iberoamericana and Vervuert, 2012)

Binkley, Peter, ed., *Pre-Modern Encyclopaedic Texts: Proceedings of the Second COMERS Congress, Groningen, 1–4 July 1996* (Leiden: Brill, 1997)

Blake, N. F., *Caxton: England's First Publisher* (London: Barnes and Noble, 1976)

Blake, N. F., *William Caxton and English Literary Culture* (London: Hambledon, 1991)

Bottici, Chiara, and Benoît Challand, eds, *Imagining Europe: Myth, Memory, and Identity* (Cambridge: Cambridge University Press, 2014)

Boucher, Caroline, 'De la *subtilité* en français: vulgarisation et savoir dans les traductions d'*auctoritates* des XIIIe–XIVe siècles', in *The Theory and Practice of Translation in the Middle Ages*, ed. by R. Voaden, R. Tixier, T. Sanchez Roura, and J. R. Rytting (Turnhout: Brepols, 2003), pp. 89–99

Bouloux, Nathalie, 'From Gaul to the Kingdom of France: Representations of French Space in the Geographical Texts of the Middle Ages (Twelfth–Fifteenth Centuries)', in *Space in the Medieval West*, ed. by Cohen and Madeline, pp. 197–217

Boyd, Matthieu, 'Celts Seen as Muslims and Muslims Seen by Celts in Medieval Literature', in *Contextualizing the Muslim Other in Medieval Christian Discourse*, ed. by Jerold C. Frakes (New York: Palgrave Macmillan, 2011), pp. 21–38

Branch, Jordan, *The Cartographic State: Maps, Territory, and the Origins of Sovereignty* (Cambridge: Cambridge University Press, 2013)

Bratsch-Prince, Dawn, 'Textual History of *Li livres dou tresor*: Fitting the Pieces Together', *Manuscripta*, 31 (1993), 276–89

Bratu, Christian, '*Translatio*, autorité et affirmation de soi chez Gaimar, Wace et Benoît de Sainte-Maure', in *The Medieval Chronicle VIII*, ed. by Erik Kooper and Sjoerd Levelt (Leiden: Brill, 2013), pp. 135–64

Braude, Benjamin, 'The Sons of Noah and the Construction of Ethnic and Geographical Identities in the Medieval and Early Modern Periods', *The William and Mary Quarterly*, 54 (1997), 103–42

Braudel, Fernand, *La Méditerranée et le Monde Méditerranéen à l'Epoque de Philippe II* (Paris: A. Colin, 1949)

Brauer, Ralph W., *Boundaries and Frontiers in Medieval Muslim Geography* (Philadelphia: American Philosophical Society, 1995)

Brennetot, Arnaud, and Muriel Rosemberg, 'Géographie de l'Europe et géographie de la construction européenne', *L'Espace Politique* (online), 19 (2013), DOI: 10.4000/espacepolitique.2613

Brincken, Anna Dorothee von den, *Fines Terrae: Die Enden der Erde und der vierte Kontinent auf mittelalterlichen Weltkarten* (Hannover: Hahnsche Buchhandlung, 1992)

Brown, Katherine A., 'The Vernacular Universe: Gossuin de Metz's *Image du monde, translatio studii*, and Vernacular Narrative', *Viator*, 44 (2013), 137–58

Brunner, Otto, *Land and Lordship: Structures of Governance in Medieval Austria*, trans. by Howard Kaminsky and James van Horn Melton (Philadelphia: University of Pennsylvania Press, 1992)

Buridant, Claude, '*Translatio medievalis*: Théorie et pratique de la traduction médiévale', *Travaux de linguistiaue et de littéature*, 21 (1983), 83–136

Burke, Peter, 'The Circulation of Knowledge', in *The Renaissance World*, ed. by J. J. Martin (New York: Routledge, 2007), pp. 191–207

Burns, Robert I., ed., *Emperor of Culture. Alfonso X the Learned of Castile and His Thirteenth-Century Renaissance* (Philadephia: University of Pennsylvania Press, 2015)

Busby, Keith, *French in Medieval Ireland, Ireland in Medieval French: The Paradox of Two Worlds* (Turnhout: Brepols, 2017)

Cachey Jr., Theodore J., 'Cosmology, geography, and Cartography', in *Dante in Context*, ed. by Zygmunt G. Barański, and Lino Pertile (Cambridge: Cambridge University Press, 2015), pp. 221–40

Cahiers d'histoire mondiale/Journal of World History/Cuadernos de historia mundial, 9.3 (1966)

Calkin, Siobhain Bly, *Saracens and the Making of English Identity: The Auchinleck Manuscript*, Studies in Medieval History and Culture (New York and Abingdon: Routledge, 2005)

Campbell, Mary B., *The Witness and the Other World: Exotic European Travel Writing, 400–1600* (Ithaca, NY and London: Cornell University Press, 1988)

Canova, Andrea, 'Altri furti boiardeschi ("Inamoramento de Orlando", II, xxviii)', *Parole rubate / Purloined Letters*, 18 (2018), 163–72; online at www.parolerubate.unipr.it (last accessed 9 December 2019)

Carmody, Francis J., 'Brunetto Latini's *Tresor*: Latin Sources on Natural Science', *Speculum*, 12 (1937), 359–66

Carmody, Francis J., 'Latin Sources of Brunetto Latini's World History', *Speculum*, 11 (1936), 359–70

Carruthers, Mary, *The Book of Memory: A Study of Memory in Medieval Culture*, 2nd edn (Cambridge: Cambridge University Press, 2008)

Cartlidge, Neil, ed., *Boundaries in Medieval Romance* (Woodbridge: Boydell & Brewer, 2008)

Cassidy, Vincent H., 'Other Fortunate Islands and Some that were Lost', *Terrae Incognitae*, 1 (1969), 35–39

Cassidy-Welch, Megan, 'Space and Place in Medieval Contexts', *Parergon*, 27 (2010), 1–12

Catlos, Brian A., and Sharon Kinoshita, eds, *Can We Talk Mediterranean? Conversations on an Emerging Field in Medieval and Early Modern Studies* (Basingstoke: Palgrave Macmillan, 2017)

Cavallo, Jo Ann, *The World Beyond Europe in the Romance Epics of Boiardo and Ariosto* (Toronto, Buffalo, and London: University of Toronto Press, 2013)

Cawsey, Kathy, 'Disorienting Orientalism: Finding Saracens in Strange Places in Late Medieval English Manuscripts', *Exemplaria*, 21 (2009), 380–97

Centili, Sara, 'La seconda redazione in versi dell'*Image du monde*: una riscrittura didactica', *Cultura neolatina*, 66 (2006), 161–206

Centili, Sara, 'La tradition manuscrite de l'*Image du monde*: Fortune et diffusion d'une encyclopédie du XIIIe siècle' (unpublished PhD thesis, École des chartes, 2005)

Chakrabarty, Dipesh, *Provincializing Europe: Postcolonial Thought and Historical Difference* (Princeton, NJ: Princeton University Press, 2000)

Chekin, Leonid S., *Northern Eurasia in Medieval Cartography: Inventory, Text, Translation, and Commentary* (Turnhout: Brepols, 2006)

Chiesa, Lorenzo, *Subjectivity and Otherness: A Philosophical Reading of Lacan* (Cambridge, MA: MIT Press, 2007)

Clarke, Dorothy Clotelle, *Juan de Mena's 'Laberinto de Fortuna': Classic Epic and 'Mester de Clerecía'*, Romance Monographs, 5 (University, MS: University of Mississippi, 1973)

Classen, Albrecht, ed., *Meeting the Foreign in the Middle Ages* (New York: Routledge, 2002)

Classen, Albrecht, review of Klaus Oschema, *Bilder von Europa im Mittelalter* (Mittelalter-Forschungen, 43) (2013), *Mediaevistik*, 27 (2014), 245–46

Clunies Ross, Margaret, 'The Influence of the Medieval Encyclopedia on Snorri's *Edda*', in *Skaldskaparmal. Snorri Sturluson's Ars Poetica and Medieval Theories of Language*, ed. by M. C. Ross, Viking Collection, 4 (Odense: Odense University Press, 1987), pp. 151–73; originally published in *The Sixth International Saga Conference 28/7–2/8 1985* (Copenhagen: Arnamagnæanske institut, 1985), pp. 177–206

Clunies Ross, Margaret, and Rudolf Simek, 'Encyclopedic Literature', in *Medieval Scandinavia: An Encyclopedia*, ed. by Philip Pulsiano, Kirsten Wolf, Paul Acker, and Donald K. Fry (New York and London: Garland, 1993), pp. 164–66

Cohen, Meredith, and Fanny Madeline, eds, *Space in the Medieval West: Places, Territories, and Imagined Geographies* (Farnham and Burlington: Ashgate, 2014)

Connochie-Bourgne, Chantal, 'Le cas de l'*Image du monde*: une encyclopédie du XIIIe siècle, ses sources antiques, l'apport médiéval', in *La transmission des connaissances techniques: tables rondes Aix-en-Provence, avril 1993–mai 1994*, edited by M.-Cl. Amouretti and G. Comet, Cahier d'histoire des techniques, 3 (Aix-en-Provence: Publications de l'Université de Provence, 1995), pp. 87–98

Connochie-Bourgne, Chantal, 'Limites et diversités de l'Europe: le parti pris par Gossouin de Metz (*Image du monde*, 1245)', in *De la Chrétienté à l'Europe. Actes du Colloque Orléans, mai 1993*, ed. by Bernard Ribémont (Orleans: Paradigme, 1995), pp. 49–62

Connochie-Bourgne, Chantal, 'L'Orient, réalité et discours, dans l'Image du monde', in *Senefiance 11: Images et signes de l'Orient dans l'Occident médiéval* (Aix-en-Provence: Presses universitaires de Provence, 1982), http://books.openedition.org/pup/2859 (accessed 23 December 2019)

Connochie-Bourgne, Chantal, 'Pourquoi et comment réécrire une encyclopédie? Deux rédactions de l'Image du monde', in *Encyclopédies médiévales: discours et savoirs*, ed. by B. Baillaud, J. de Gramont, and D. Hüe (Rennes: Presses Universitaires de Rennes et Association Diderot, 1999), pp. 143–54

Constable, Giles, 'Frontiers in the Middle Ages', in *Frontiers in the Middle Ages*, ed. by Merisalo and Pahta, pp. 3–28

Copeland, Rita, 'Language Frontiers, Literary Form, and the Encyclopedia', in *Frontiers in the Middle Ages*, ed. by Merisalo and Pahta, pp. 507–24

Cordiolani, A., 'Un manuscrit de comput et d'astronomie des XIIème–XIVème siècles, le manuscrit 467 de l'université de Glasgow de l'*Imago mundi*', *Scriptorium*, 3.1 (1949), 75–79

Cornish, Alison, *Vernacular Translation in Dante's Italy: Illiterate Literature* (Cambridge: Cambridge University Press, 2011)

Couzin, Robert, *Right and Left in Early Christian and Medieval Art* (Leiden and Boston: Brill, 2021)

Crick, Julia, 'The Power and the Glory: Conquest and Cosmology in Edwardian Wales (Exeter, Cathedral Library, 3514', in *Textual Cultures, Cultural Texts*, ed. by Orietta da Rold and Elaine Treharne, Essays and Studies, 63 (Cambridge: Brewer, 2010), pp. 21–42

Crosby, Ruth, 'Oral Delivery in the Middle Ages', *Speculum*, 11 (1936), 88–110

Crouse, Robert Darwin, 'Honorius Augustodunensis: Disciple of Anselm?', *Analecta Anselmiana*, 4 (1975), 131–39

D'Amico, Juan Carlos, 'Fazio degli Uberti, il Dittamondo e l'ideologia imperiale', in *Emperors and Imperial Discourse in Italy, c. 1300–1500: New Perspectives*, ed. by Anne Huijbers (Rome: École Française de Rome, 2022), pp. 89–104

Dainotto, Roberto M., *Europe (In Theory)* (Durham and London: Durham University Press, 2007)

Dalché, Patrick Gautier, 'Représentations géographiques de l'Europe – septentrionale, centrale et orientale – au Moyen Age', in *Europa im Weltbild des Mittelalters: kartographische Konzepte*, ed. by Ingrid Baumgärtner and H. Kugler, Orbis Mediaevalis: Vorstellungswelten des Mittelalters, 10 (Berlin: Akademie, 2008), pp. 63–79

Dalché, Patrick Gautier, *La géographie de Ptolémée en Occident (IVe–XVIe siècle)* (Turnhout: Brepols, 2009)

Dalché, Patrick Gautier, '*Hic mappa mundi considerata est*: lecture de la mappemonde au Moyen Age', in *Itinerari del testo per Stefano Pittaluga*, ed. by C. Cocco, Clara Fossati, Attilio Grisafi, Francesco Mosetti Casaretto, and Giada Boiani, Publicazzioni del D.AR.FI.CL.ET. 'Francesco Della Corte', Third Series, 254 (Genova: Dipartimento di Antichità, Filosofia e Storia, 2018), pp. 495–516

Dalché, Patrick Gautier, 'De la glose à la contemplation: place et function de la carte dans les manuscrits de haut Moyen Âge', in *Testo e immagine nell'alto medioevo: 15–21 aprile 1993*, Settimane di studio del Centro Italiano di Studi sull'alto Medioevo, 41 (Spoleto: Presso la sede del Centro, 1994), pp. 693–771

Dalché, Patrick Gautier, 'Maps in Words: The Descriptive Logic of Medieval Geography from the Eighth to the Twelfth Century', in *The Hereford World Map: Medieval World Maps and their Context*, ed. by P. D. A. Harvey (London: BL, 2006), pp. 223–42

Dalché, Patrick Gautier, 'The Reception of Ptolemy's *Geography* (End of the Fourteenth to Beginning of the Sixteenth Century' in *The History of Cartography, Vol. 3: Cartography in the European Renaissance, Part 1*, ed. by David Woodward (Chicago: University of Chicago Press, 2007), pp. 285–364

Davies, Surekha, 'The Wondrous East in the Renaissance Geographical Imagination: Marco Polo, Fra Mauro and Giovanni Battista Ramusio', *History and Anthropology*, 23 (2012), 215–34

Dawson, Christopher, *Understanding Europe* (Washington, DC: Catholic University of America Press, 2008; first published London: Sheed and Ward, 1952)

Dębicki, Jacek, *Zachodni portal katedry świętego Łazarza w Autun: Studium z historii sztuki i historii idei* (Kraków: Universitas 2002)

Degering, Hermann, *Kurzes Verzeichnis der germanischen Handschriften der Preussischen Staatsbibliothek I: Die handschriften in Folioformat* (Leipzig: Verlag von Karl W. Hiersemann, 1925)

Delanty, Gerard, *Formations of European Modernity: A Historical and Political Sociology of Europe* (London: Palgrave Macmillan, 2019)

Delanty, Gerard, *Inventing Europe: Idea, Identity, Reality* (Basingstoke: Macmillan 1995)

Delivré, Fabrice, 'Du nouveau sur la "Liste de Florence": La chronique du Pseudo-Godel (v. 1175) et la prehistoire du Provinciale Romanum du XIIIe siècle', *Bibliothèque de l'École des chartes*, 167 (2009), 353–74

Destombes, Marcel, *Mappaemondes A.D. 1200–1500: Catalogue préparé par la commission des cartes anciennes de l'Union géographique internationale* (Amsterdam: N. Israel, 1964)

Deyermond, Alan, 'Building a World: Geography and Cosmology in Castilian Literature of the Early Thirteenth Century', *Canadian Review of Comparative Literature*, 23 (1996), 141–59

Díaz Fernández, Alejandro, ed., *Provinces and Provincial Command in Republican Rome: Genesis, Development and Governance* (Sevilla and Zaragoza: Editorial Universidad de Sevilla and Prensas de la Universidad de Zaragoza, 2021)

Dion, Elisée, 'Penser le monde en langue vulgaire et en images: enjeux sociaux et dimensions visuelles de la transmission du savoir dans l'Image du monde (XIIIe–XVe siècles)' (unpublished PhD thesis, Université de Reims and Eberhard-Karls-Universität, 2018)

Doberentz, O., 'Die Erd- und Völkerkunde in de Weltchronik des Rudolf von Hohen-Ems', *Zeitschrift für deutsche Philologie*, 12 (1881), 387–454

Doberentz, O., 'Die Erd- und Völkerkunde in de Weltchronik des Rudolf von Hohen-Ems (*Fortsetzung*)', *Zeitschrift für deutsche Philologie*, 13 (1882), 29–56

Domínguez, Frank A., 'Juan de Mena's *El Laberinto de Fotruna*, Petrarch's Africa, the Scipio/Caesar Controversy', *La Crónica*, 42 (2014), 137–68

Domínguez Rodríguez, Ana, *Astrología y arte en el 'Lapidario' de Alfonso X el Sabio* (Madrid: Edilán, 1984)

Donkin, Lucy E. G., '"*Usque ad ultimum terrae*": Mapping the Ends of the Earth in Two Medieval Floor Mosaics', in *Cartography in Antiquity and the Middle Ages*, ed. by Talbert and Unger, pp. 189–217

Draelants, Isabelle, 'Le "siècle de l'encyclopédisme": conditions et critères de définition d'un genre', in *Encyclopédire: Formes de l'ambition encyclopédique dans l'Antiquité et au Moyen Âge*, ed. by Arnaud Zucker, Collection d'études médiévales de Nice, 14 (Tunrhout: Brepos, 2013), pp. 81–106

Ducos, Joëlle, 'Que traduire en français? Traductions uniques et traductions multiples', in *Translation and Authority: Authorities in Translation*, ed. by Pieter de Leemans and Michèle Goyens, The Medieval Translator, 16 (Turnhout: Brepols, 2017), pp. 39–52

Duhem, P., *Le Système du monde, histoire des doctrines cosmologiques de Platon à Copernic*, vol. III (Paris: Hermann, 1919)

Dunphy, R. Graeme, 'Historical Writing in and after the Old High German Period', in *German Literature of the Early Middle Ages*, ed. by Brian Murdoch (Woodbridge and New York: Camden House and Boydell & Brewer, 2004), pp. 201–26

Dworkin, Steven N., *A Guide to Old Spanish* (Oxford: Oxford University Press, 2018)

Eckhardt, Caroline D., 'One Third of the Earth? Europe Seen and Unseen in the Middle English Chronicles of the Fourteenth Century', *Comparative Literature*, 58.4, 'The Idea of Europe' (2006), 313–38

Eco, Umberto, *Mouse or Rat? Translation as Negotiation* (London: Weidenfeld & Nicholson, 2003)

Edson, Evelyn, *Mapping Time and Space: How Medieval Mapmakers Viewed their World*, British Library Studies in Map History, 1 (London: BL, 1997)

Edson, Evelyn, 'Maps in Context: Isidore, Orosius, and the Medieval Image of the World', in *Cartography in Antiquity and the Middle Ages*, ed. by Talbert and Unger, pp. 21–236

Edson, Evelyn, 'The Medieval World View: Contemplating the Mappamundi', *History Compass*, 8 (2010), 503–17, doi/10.111/j.1478-0542.2010.00676.x

Edson, Evelyn, *The World Map, 1300–1492. The Persistence of Tradition and Transformation* (Baltimore: Johns Hopkins University Press, 2007)

Egeler, Matthias, 'Eikþyrnir and the Rivers of Paradise: Cosmological Perspectives on Dating *Grímnismál* 26–28', *Arkiv för nordisk filologi*, 128 (2013), 17–39

Ellenblum, Ronnie, 'Were there Borders and Borderlines in the Middle Ages? The Example of the Latin Kingdom of Jerusalem', in *Medieval Frontiers*, ed. by Abulafia and Berend, pp. 105–20

Ellis, Roger, 'Translation and Frontiers in Late Medieval England: Caxton, Kempe, and Mandeville', in *Frontiers in the Middle Ages*, ed. by Merisalo and Pahta, pp. 559–83

Engel, William E., Rory Loughnane, and Grant Williams, eds, *The Memory Arts in Renaissance England: A Critical Anthology* (Cambridge: Cambridge University Press, 2016)

Falk, Seb, *The Light Ages: A Medieval Journey of Discovery* (Milton Keynes: Allen Lane, 2020)

Falmagne, T., D. Stutzmann, A.-M. Turcan-Verkerk, and P. Gandil, eds, *Les cisterciens et la transmission des textes (XIIIe–XVIIIe siècles)* (Turnhout: Brepols, 2018)

Fawtier, Robert, 'Comment le roi de France, au début du XIVe siècle, pouvait-il se représenter son royaume', in *Mélanges offerts à M. Paul-Edmond Martin* (Geneva: Comité des mélanges P.-E. Martin, 1961), pp. 65–77

Febvre, Lucien, 'Un livre qui grandit: *La Médieranée et le monde méditeranéen a l'époque de Philippe II*', *Revue historique*, 203.2 (1950), 216–224

Firchow, Evelyn Scherabon, and Kaaren Grimstad, eds, *'Elucidarius' in Old Norse Translation* (Reykjavik: Stofnun Árna Magnússonar, 1989)

Flanigan, C. Clifford, 'The Moving Subject: Medieval Liturgical Processions in Semiotic and Cultural Perspective', in *Moving Subjects: Processional Performance in the Middle Ages and the Renaissance*, ed. by Kathleen Ashley and Wim Hüsken (Amsterdam: Rodopi, 2001), pp. 35–52

Flint, Valerie I. J., 'Honorius Augustodunensis', in *Authors of the Middle Ages II: Historical and Religious Writers of the Latin West*, nos 5–6, ed. by Patrick J. Geary (Aldershot: Variorum Reprints, 1995), pp. 89–183

Flint, Valerie I. J., *The Imaginative Landscape of Christopher Columbus* (Princeton, NJ: Princeton University Press, 1992)

Flint, Valerie I. J., 'World History in the Early Twelfth Century: The *Imago mundi* of Honorius Augustodunensis', in *The Writing of History in the Middle Ages: Essays Presented to Richard William Southern*, ed. by R. H. C. Davis, and J. M. Wallace-Hadrill (Oxford: Clarendon Press, 1981), pp. 211–38

Fowler, Robert L., 'Encyclopaedias: Definitions and Theoretical Problems', in *Pre-Modern Encyclopaedic Texts*, ed. by Binkley, pp. 3–29

Frakes, J. C., ed., *Contextualizing the Muslim Other in Medieval Christian Discourse*, The New Middle Ages (New York: Palgrave Macmillan, 2011)

Franklin-Brown, Mary, *Reading the World: Encyclopedic Writing in the Scholastic Age* (Chicago: Chicago University Press, 2012)

Friedman, John Block, 'Cultural Conflicts in Medieval World Maps', *Implicit Understandings: Observing, Reporting, and Reflecting on the Encounters between Europeans and Other Peoples in the Early Modern Era*, edited by Stuart B. Schwartz (Cambridge: Cambridge University Press, 1994), pp. 64–95

Fromherz, Allen James, *The Near West: Medieval North Africa, Latin Europe and the Mediterranean in the Second Axial Age* (Edinburgh: Edinburgh University Press, 2016)

Gaida, Margaret, 'Reading *Cosmographia*: Peter Apian's Book-Instrument Hybrid and the Rise of the Mathematical Amateur in the Sixteenth Century', *Early Science and Medicine*, 21 (2016), 277–302

Garrigues, Marie-Odile, 'Honorius Augustodunensis et L'Italie', *Mélanges de l'École Française de Rome. Moyen Âge, Temps Modernes*, 84 (1952), 511–30

Garrigues, Marie-Odile, 'L'oeuvre d'Honorius Augustodunensis: Inventaire Critique' (1), *Abhandlungen der Braunschweigischen Wissenschaftlichen Gesellschaft*, 38 (1986), 7–136

Gaunt, Simon, *Marco Polo's 'Le Devisement du Monde': Narrative Voice, Language and Diversity* (Cambridge: D. S. Brewer, 2013)

Geary, Patrick J., *The Myth of Nations* (Princeton, NJ: Princeton University Press, 2002)

Goetz, Hans-Werner, '*Gens*: Terminology and Perception of the "Germanic" Peoples from Late Antiquity to the Early Middle Ages', in *The Construction of Communities in the Early Middle*

Ages: Texts, Resources, and Artefacts, ed. by R. Corrandini, M. Diesenberger, and Helmut Reimitz (Leiden and Boston: Brill, 2003), pp. 39–64

Goldie, Matthew Boyd, *Scribes of Space: Place in Middle English Literature and Late Medieval Science* (Ithaca, NY and London: Cornell University Press, 2019)

Gottschall, Dagmar, 'Marius Salernitanus und Gottfried von Viterbo', *Sudhoffs Archiv*, 75 (1991), 111–13

Gregory, Tullio, 'Translatio Studiorum', in *Translatio Studiorum: Ancient, Medieval and Modern Bearers of Intellectual History*, ed. by M. Sgarbi (Leiden: Brill, 2014), pp. 1–22

Green, D. H., *Women Readers in the Middle Ages* (Cambridge: Cambridge University Press, 2007)

Gunnar Harðarson, 'Old Norse Intellectual Culture: Appropriation and Innovation', in *Intellectual Culture in Medieval Scandinavia, c. 1100–1350*, ed. by Stefka Georgieva Eriksen (Turnhout: Brepols, 2016), pp. 35–73

Gybbon-Monypenny, G. B., 'The Spanish *Mester de Clerecía* and its Intended Public: Concerning the Validity as Evidence of Passages of Direct Address to the Audience', in *Medieval Miscellany Presented to Eugène Vinaver*, ed. by F. Whitehead, A. H. Diverres, and F. E. Sutcliffe (Manchester: Manchester University Press, 1965), pp. 230–44

Halmos, Paul R., *Naïve Set Theory* (Princeton, NJ: Princeton University Press, 1960)

Halvorsen, Eyvindr Fjeld, 'Landfreðgatal', *Kulturhistorische leksikon for nordisk middelalder*, 10 (1965), cols. 311–13

Hamilton, George L., 'Encore un plagiat médiéval: la *Mappemonde* de Pierre de Beauvais et les *Divisiones mundi* de Perot de Garbelai', in *Mélanges de linguistique et de littérature offerts à M. Alfred Jeanroy par ses élèves et ses amis* (Paris: Droz, 1928), pp. 627–38

Hannam, Walter Andrew, 'The *Ineuitabile* of Honorius Augustodunensis: A Study in the Textures of Early Twelfth-Century Augustinianisms' (unpublished PhD dissertation, Boston College, Graduate School of Arts and Sciences, 2013)

Harley, J. B., 'The Map and the Development of the History of Cartography', in *History of Cartography I*, ed. by Harley and Woodward, pp. 1–42

Harley, J. B., 'Maps, Knowledge, and Power', in *The Iconography of Landscape: Essays on the Symbolic Representation, Design and Use of Past Environments*, ed. by Denis Cosgrove and Stephen Daniels (Cambridge: Cambridge University Press, 1988), pp. 277–312

Harley, J. B., 'Silences and Secrecy: The Hidden Agenda of Cartography in Early Modern Europe', *Imago mundi*, 40 (1988), 57–76

Harley, J. B., and David Woodward, eds, *The History of Cartography I: Cartography in Prehistoric, Ancient, and Medieval Europe and the Mediterranean*, (Chicago: Chicago University Press, 1987)

Harvey, P. D. A., ed., *The Hereford World Map: Medieval World Maps and their Context* (London: BL, 2006)

Harvey, P. D. A., *Medieval Maps* (Toronto: University of Toronto Press, 1991)

Harvey, P. D. A., 'Medieval Maps: An Introduction' in *History of Cartography I*, ed. by Harley and Woodward, pp. 283–85

Harvey, Steven, ed., *The Medieval Hebrew Encyclopedias of Science and Philosophy: Proceedings of the Bar-Ilan University Conference*, Amsterdam Studies in Jewish Thought, 7 (Dordrecht: Springer Science + Business Media, 2000)

Hawking, Stephen, *A Brief History of Time. From the Big Bang to Black Holes* (London: Bantam, 2016)

Hay, Denys, *Europe: The Emergence of an Idea* (Edinburgh: Edinburgh University Press, 1957)

Hay, Denys, *The Medieval Centuries*, 2nd edn (London: Methuen, 1964)

Heather, Peter J., *The Restoration of Rome: Barbarian Popes and Imperial Pretenders* (Oxford: Oxford University Press, 2013)

Heffernan, Michael, 'The Changing Political Map: Geography, Geopolitics, and the Idea of Europe since 1500', in *An Historical Geography of Europe*, ed. by R. A. Butlin and R. A. Dodgshon (Oxford: Clarendon Press, 1998), pp. 140–80

Heffernan, Michael, *The European Geographical Imagination* (Stuttgart: Franze Steiner, 2006)

Hellinga, Lotte, *William Caxton and Early Printing in England* (London: BL, 2010)

Hen, Yitzhak, 'The Church in Sixth-Century Gaul', in *A Companion to Gregory of Tours*, ed. by Alexander C. Murray (Leiden and Boston: Brill, 2015), pp. 232–55

Herbers, Klaus, 'Europa und seine Grenzen im Mittelalter', in *Grenzräume und Grenzüberschreitungen im Vergleich. Der Osten und der Westen des mittelalterlichen Lateineuropa*, ed. by Klaus Herbers and Nikolas Jaspert, Europa im Mittelalter, 7 (Berlin: Akademie Verlag, 2007), pp. 21–41

Herbers, Klaus, review of Klaus Oschema, *Bilder von Europa im Mittelalter* (Mittelalter-Forschungen, 43) (2013), *Zeitschrift für Historische Forschung*, 43 (2016), 77–78

Hiatt, Alfred, 'Maps of Empires Past', in *Post-Empire Imaginaries? Anglophone Literature, History, and the Demise of Empires*, ed. by Barbara Buchenau, Virginia Richter and Marijke Denger (Leiden and Boston: Brill, 2015), pp. 3–23

Hill, Thomas D., '*Rígsþula*: Some Medieval Christian Analogues', *Speculum*, 61 (1986), 79–89

Hobsbawm, Eric, 'The Curious History of Europe', in *On History* (London: Weidenfield & Nicholson, 1997), pp. 217–27

Hoogvliet, Margriet, '*Mappae mundi* and Medieval Encyclopaedias: Image versus Text', in *Pre-Modern Encyclopaedic Texts*, ed. by Binkley, pp. 63–74

Hoogvliet, Margriet, *Pictura et scriptura: textes, images et herméneutique des mappae mundi (XIIIe–XVIe siècle)* (Turnhout: Brepols, 2007)

Isaac, Benjamin, *The Limits of Empire: The Roman Army in the East* (Oxford: Clarendon Press, 1990)

Italiano, Federico, *Translation and Geography* (Abingdon: Routledge, 2016)

Iwańczak, Wojciech, 'Borders and Borderlines in Medieval Cartography', in *Frontiers in the Middle Ages*, ed. by Merisalo and Pahta, pp. 661–72

Jakobsen, Johnny Grandjean Gøgsig, 'Friars of Science: Dominican Transmission and Usage of Scientific Knowledge in Medieval Scandinavia' in *Medieval Science in the North: Travelling Wisdom, 1000–1500*, ed. by Christian Etheridge and Michele Campopiano (Turnhout: Brepols, 2021), pp. 141–58

Jamróz, Marek, *Bóg, świat i człowiek w pismach Honoriusza Augustodunensisa* (Lublin: Wydawnictwo KUL 2008)

Jensen, Helle, '*Eiríks saga víðförla*', in ed., *Medieval Scandinavia: An Encyclopedia*, ed. by Philip Pulsiano, Kirsten Wolf, Paul Acker, and Donald K. Fry (New York and London: Garland, 1993), pp. 160–61

Jónas Kristjánsson, 'Elements of Learning and Chivalry in *Fóstbræðra saga*', in *The Icelandic Sagas and Western Literary Tradition. Proceedings of the First International Saga Conference, University of Edinburgh, 21–29 August, 1971*, ed. by Hermann Pálsson and Desmond Slay (London: Viking Society for Northern Research, 1973), pp. 259–99

Jones, Chris, 'Understanding Political Conceptions in the Later Middle Ages: The French Imperial Candidatures and the Idea of the Nation-State', *Viator*, 42 (2011), 83–114

Jongkees, A. G., 'Translatio Studii: les avatars d'une thème médiéval', *Miscellanea Medievalia in Memoriam Jan Frederick Niermeyer* (Groningen: J. B. Wolters, 1967), pp. 41–51

Jordan, William Chester, '"Europe" in the Middle Ages', in *The Idea of Europe*, ed. by Pagden, pp. 72–90

Jostkleigrewe, Georg, 'L'espace entre tradition et innovation : La géographie symbolique du monde et son adaptation par Gossouin de Metz', *Construction de l'espace au Moyen Age:*

Pratiques et représentations. XXXVIIe Congrès de la SHMES. Mulhouse, 2–4 juin 2006 (Paris: Publications de la Sorbonne, 2007), 369–78

Kedwards, Dale, 'Geography', in *A Critical Companion to Old Norse Literary Genre*, ed. by Massimiliano Bampi, Carolyne Larrington, and Sif Rikharðsdóttir (Cambridge: D. S. Brewer, 2020), pp. 127–44

Kedwards, Dale *The Mappae Mundi of Medieval Iceland* (Woodbridge: Boydell & Brewer, 2020)

Kenny, Neil, *The Palace of Secrets: Béroalde de Verville and Renaissance Conceptions of Knowledge* (Oxford: Oxford University Press, 1991)

Kerkhof, Maxim P. A. M., 'Las glosas al *Laberinto de Fortuna* en los mss. PN7 y PMM1 (*olim* MM1)', in *Juan de Mena: De letrado a poeta*, ed. by Cristina Moya García (Cambridge: Boydell and Brewer, 2015), pp. 217–54

Kerkhof, Maxim P. A. M., 'Sobre las edciones del *Laberinto de Fortuna* publicadas de 1481 a 1943, y la tradición manuscrita', in *Forum Litterarum: Miscelanea de estudos literários, linguisticos e históricos oferecida a J. J. van den Besselaar*, ed. by J. A. H. Bots and Maxim P. A. M. Kerkhof (Amsterdam and Maarssen: APA and Holland University Press, 1984), pp. 26–82

Kerkhof, Maxim P. A. M., and Rob Le Pair, 'El *Laberinto de Fortuna* de Juan de Mena: las ediciones en relación con la tradición manuscrita', *Homenaje al Profesor Antonio Vilanova*, I, ed. by Marta Cristina Carbonell and Adolfo Sotelo Vázquez (Barcelona: Universidad de Barcelona, 1989), pp. 321–39

Kinkade, Richard P., ed., *Los Lucidarios españoles. Estudio y edición* (Madrid: Gredos, 1968)

Kinkade, Richard P., 'Un nuevo manuscrito de la *Semeiança del mundo*', *Hispanic Review*, 39 (1971), 261–70

Kinoshita, Sharon, 'Discrepant Medievalisms: Deprovincializing the Middle Ages', in *Worldlings: World Literature, Field Imaginaries, Future Practices: Doing Cultural Studies in the Era of Globalization*, ed. by R. Wilson and Chris Connery (Santa Cruz: New Pacific Press, 2006), pp. 75–89

Kinoshita, Sharon, *Medieval Boundaries: Rethinking Difference in Old French Literature* (Philadelphia, PA: University of Pennsylvania Press, 2006)

Kirby, Ian J., *Bible Translation in Old Norse* (Geneva: Droz, 1986)

Klebel, Ernst, 'Das Fortleben des Namens "Noricum" im Mittelalter', *Carinthia I. Mitteilungen des Geschichtsvereins für Kärnten, Klagenfurt*, 146 (1956), 481–92

Kleinhans, Martha, *'Lucidere vault tant a dire comme donnant lumiere': Untersuchung und Edition der Prosaversionen 2, 4 und 5 des 'Elucidarium'*, Beihefte zur Zeitschrift für romanische Philologie, 248 (Tübingen: Niemeyer, 1993)

Kleivane, Elise, 'There is More to Stjórn than Biblical Translation', in *Speculum septentrionale: Konungs skuggsjá and the European Encyclopedia of the Middle Ages*, ed. by Karl G. Johansson and Elise Kleivane (Oslo: Novus Forlag, 2018), pp. 115–47

Kline, Naomi Reed, *Maps of Medieval Thought: The Hereford Paradigm* (Woodbridge: Boydell, 2001)

Knapp, James A., 'Translating for Print: Continuity and Change in Caxton's *Mirrour of the World*', *Disputatio*, 3 (1998), 64–90

Kogman-Appel, Katrin, *Catalan Maps and Jewish Books: The Intellectual Profile of Elisha ben Abraham Cresques (1325–1387)* (Turnhout: Brepols, 2020)

Kohl, Thomas, 'Territorialisierungen im 11. Jahrhundert? Der deutsche Südwesten und das westliche Frankreich im Vergleich', in *Territorium. Raum und Politik: Wahrnehmung und Praxis im Frankenreich und seinen Nachfolgereichen vom 9. bis zum 13. Jahrhundert* (Tübingen, 2011)

Kroesen, Jacoba M. C., *Over de compositie der Fóstbrœðra saga* (Leiden: Universitaire Pers, 1962)

Kupfer, Marcia, 'The Noachide Dispersion in English Mappae Mundi, ca. 960–ca. 1130', *Peregrinations: Journal of Medieval Art and Architecture*, 4 (2013), 81–106

Lacan, Jacques, *Livre XVI: D'un Autre à l'autre, 1968–1969*, ed. Jacques-Alain Miller (Paris: Éditions du Seuil, 2006)

Lacomba, Marta, 'La articulación de un discurso científico en castellano bajo Alfonso X. Se l *Semeiança del mundo* al *Lapidatio*: de una descripción del mundo en romance a la construcción de un espacio castellano del saber', in *Lenguas, reinos y dialectos en la Edad Media Ibérica: la construcción de la identidad: homenaje a Juán Ramón Lodares* (Madrid and Frankfurt am Main: Iberoamericana and Vervuers, 2008)

Latella, Fortunata, 'Come lavorava un intelettuale laico del medioevo: Gli *Otia imperialia* di Gervasio di Tilbury tra *intenvio* e *compilatio*', *Revista de literatura medieval*, 25 (2013), 103–34

Lavezzo, Kathy, *Angels on the Edge of the World: Geography, Literature, and English Community, 1000–1534* (Ithaca, NY: Cornell University Press, 2006)

Lefebvre, Henri, *La production d'espace* (Paris: Anthropos, 1974)

Lefebvre, Henri, *The Production of Space*, trans. by Donald Nicholson-Smith (Malden, MA and Oxford: Blackwell, 1991)

Lefèvre, Yves, *'L'Elucidarium' et les lucidaires: Contribution, par l'histoire d'un texte, à l'histoire des croyances religieuses en France au Moyen Âge*, Bibliothèque des Écoles françaises d'Athènes et de Rome, 180 (Paris: E. de Brocard, 1954)

Le Goff, Jacques, *L'Europe est-elle née au Moyen Age?* (Paris: Seuil, 2003)

Le Goff, Jacques, *The Birth of Europe* (Malden, MA: Blackwell, 2005)

Le Goff, Jacques, 'Pourquoi le XIIIe siècle a-t-il été un siècle d'encyclopédisme?' in *L'enciclopedismo medievale*, ed. by Picone, pp. 23–40

Lehmann, Paul, 'Skandinaviens Anteil an der Lateinischen Literatur und Wissenschaft des Mittelalters', in *Erforschung des Mittelalters. Ausgewählte Abhandlungen und Ausätze*, by Paul Lehmann, 5 vols (Stuttgart: Anton Hiersemann, 1959–1973), V (1962), pp. 275–393

Leveau, Philippe '*Territorium urbis*. Le territoire de la cité romqine et ses divisions: du vocabulaire aux réalités administratives', *Revue des Études Anciennes*, 95 (1993), 459–71

Lévinas, Emmanuel, *Totalité et infini: Essai sur l'extériorité* (The Hague: Nijhoff, 1971)

Lewis, Charlton T., and Charles Short, *A Latin Dictionary* (Oxford: Clarendon Press, 1879)

Lewis, Martin W., and Karen E. Wigen, *The Myth of Continents: A Critique of Metageography* (Berkeley: University of California Press, 1997)

Leyser, Karl J., 'Concepts of Europe in the Early Middle Ages', *Past & Present*, 137 (1992), 25–47

Librandi, Rita, 'Ristoro, Brunetto, Bencivenni e la *Metaura*: intrecci di glosse e rinvii tra le opere di uno scaffale scientifico', in *Lo scaffale della biblioteca scientifica in volgare (secoli XIII–XVI)*, ed. by R. Librandi and R. Piro (Florence: Sismel–Edizioni del Galluzzo, 2006), pp. 101–22

Lida de Malkiel, María Rosa, *Juan de Mena, poeta del prerrenacimiento español* (Mexico: Publicaciones de la Nueva Revista de Filología Hispánica, 1950)

Lilley, Keith, 'Geography's Medieval History: A Neglected Enterprise?', *Dialogues in Human Geography*, 1 (2011), 147–62

Lilley, Keith, 'Introduction', in *Mappimg Medieval Geographies: Geographical Encounters in the Latin West and Beyond, 300–1600*, ed. by Keith Lilley (Cambridge: Cambridge University Press, 2013), pp. 1–20

Linehan, Peter and Janet L. Nelson, 'Introduction' to *The Medieval World*, ed. by Linehan and Nelson (London and New York: Routledge, 2001), pp. 7–13

Llompart i Moragues, Gabriel, Ramón J. Pujades, and Julio Samsó, eds, *El Atlas Catalan de Cresques Abraham* (Barcelona: Diafora, 1975)

Llompart i Moragues, Gabriel, Ramon Pujades i Bataller, and Julio Samso Moya, eds, *El mon i els dies: L'Atles Català* (Barcelona: Enciclopedia Catalana, 2005)

Lozovsky, N., *'The Earth is Our Book': Geographical Knowledge in the Latin West, ca. 400–1000* (Ann Arbor: University of Michigan Press, 2000)

Luff, Robert, *Wissensvermittlung im Europäischen Mittelalter: 'Imago mundi'-Werke und ihre Prologe* (Tübingen: Niemeyer, 1999)

Luft, Diana, Peter Wynn Thomas, and D. Mark Smith, eds, *Rhyddiaith Gymraeg 1300–1425* (2013), http://www.rhyddiaithganoloesol.caerdydd.ac.uk (last accessed 4 July 2024)

Lusignan, S., 'La topique de la *translatio studii* et les traductions françaises de textes savants au XIVe siècle', in *Traduction et traducteurs au Moyen Âge: Actes du colloque international du CNRS organisé à Paris. Institut de recherche et d'histoire des textes les 26–28 mai 1986* (Paris, 1989), pp. 303–15

Machan, Tim William, 'Chaucer as Translator', in *The Medieval Translator: The Theory and Practice of Translation in the Middle Ages. Papers Read at a Conference Held 20–23 August 1987 at the University of Wales Conference Centre, Gregynog Hall*, ed. by Roger Ellis, with Jocelyn Price, Stephen Medcalf, and Peter Meredith (Cambridge: D. S. Brewer, 1989), pp. 55–67

Martínez, Marcos, 'El mito de la isla perdida y su tradición en la historia, cartografiía, literatura y arte', *Revista de filologia de la Universidad de La Laguna*, 16 (1998), 143–84

Martínez Hernández, Marcos, 'Las Islas Afortunadas en la Edad media', *Cuadernos del CEMyR*, 14 (2006) 55–78

Mauntel, Christoph, *Die Erdteile in der Weltordnung des Mittelalters: Asien – Europa – Afrika* (Stuttgart: Anton Hiersemann Verlag, 2023)

Maurice, Jean, 'Brunetto Latini compilateur: "Deviser la nature des animaus" dans le *Trésor*', *Cahiers de recherches médiévales et humanistes*, 23 (2012), 173–89

McCluskey, Stephen C., *Astronomies and Cultures in Early Medieval Europe* (Cambridge: Cambridge University Press, 1998)

McCormick, Michael, *Origins of the European Economy: Communications and Commerce AD 300–900* (Cambridge: Cambridge University Press, 2001)

McCusker, Maeve, and Anthony Soares, eds, *Islanded Identities: Constructions of Postcolonial Cultural Insularity* (Amsterdam and New York: Rodopi, 2011)

McDougall, David 'A Note on the Shape of Rome in the *Fóstbrœðra saga*', *Gripla*, 6 (1984), 260–63

Medeiros, François de, *L'Occident et l'Afrique (XIIIe–XVe siècle): Images et représentations* (Paris: Éditions Karthala and Centre de Recherches Africaines, 1985)

Meier-Staubach, Christel, 'Organisation of Knowledge and Encyclopaedic Ordo: Functions and Purposes of a Universal Literary Genre', in *Pre-Modern Encyclopaedic Texts*, ed. by Binkley, pp. 101–80

Melamed, Abraham, 'The Hebrew Encyclopedias of the Renaissance' in *The Medieval Hebrew Encyclopedias of Science and Philosophy*, ed. by Harvey, pp. 441–64

Menegaldo, Silvère, 'Géographie et imaginaire insulaire au Moyen Âge, d'Isidore de Séville à Jean de Mandeville', *Les Lettres romanes*, 66 (2012), 37–86

Merisalo, O., and P. Pahta, eds, *Frontiers in the Middle Ages: Proceedings of the Third European Congress of Medieval Studies (Jyväskylä, 10–14 June 2003)*, Textes et Etudes du Moyen Age, 35 (Louvain-la-neuve: Féderation Internationale des Instituts d'Études Médiévales, 2006)

Merrills, A. H., *History and Geography in Late Antiquity* (Cambridge: Cambridge University Press, 2005)

Meulengracht-Sørensen, Preben, 'Modernitet og tratitionalisem: et bidrag til islændingesagaernes litteraturhistorie med en diskussion af *Fóstbrœðra sagas* alder', in *Die Aktualität der Saga: Festschrift für Hans Schottmann*, ed. by Stig Toftgaard Andersen (Berlin: De Gruyter, 1999), pp. 149–62

Meyer, Christian, 'Musique et astronomie dans le *Liber quatuor distinctionum* de Michael Scot', *Archives d'histoire doctrinale et littéraire du Moyen Âge*, 76 (2009), 119–77 DOI: 10.3917/ahdlm.076.0119

Meyer, Paul, '*L'Image du monde*, rédaction du MS. Harley 4333', *Romania*, 21 (1892), 481–505

Mikkeli, Heikki, *Europe as an Idea and an Identity* (New York, 1998)

Miller, James A., review of Ralph W. Brauer, *Boundaries and Frontiers in Medieval Muslim Geography* (1995), *International Journal of Middle East Studies*, 29 (1997), 439–41

Minnis, Alistair J., *Translations of Authority in Medieval English Literature: Valuing the Vernacular* (Cambridge: Cambridge University Press, 2009)

Mittman, Asa Simon, 'The Other Close at Hand: Gerald of Wales and the "Marvels of the West"', in *The Monstrous Middle Ages*, ed. by Bettina Bildhauer and Robert Mills (Toronto: University of Toronto Press, 2003), pp. 97–112

Mollat du Jourdin, Michel, *Europe and the Sea* (Oxford: Blackwell, 1993)

Monmonier, Mark S., *Drawing the Line: Tales of Maps and Cartocontroversy* (New York: Henry Holt and Company, 1995)

Monmonier, Mark S., *Rhumb Lines and Map Wars: A Social History of the Mercator Projection* (Chicago and London: University of Chicago Press, 2010)

Montero, Ana M., 'El *Lucidario* de Sancho IV: redefinición de su relación textual con el *Elucidarius* de Honorius Augustodunensis y el *Setenario* de Alfonso X', in *The Theory and Practice of Translation in the Middle Ages*, ed. by R. Voaden, R. Tixier, T. Sanchez Roura and J. R. Rytting (Turnhout: Brepols, 2003), pp. 49–59

Moya García, Cristina, 'Juan de Mena, Álvaro de luna y los Mendoza: Literatura y estrategias de linaje', in *Juan de Mena: De letrado a poeta*, ed. by Moya García, pp. 55–74

Moya García, Cristina, ed., *Juan de Mena: De letrado a poeta*, (Cambridge: Boydell and Brewer, 2015)

Musikas, Claire, 'A Road to India: the example of *Eireks saga víðförla*', *Brathair*, 11 (2011), 95–104

Napolitano, David, 'Adjusting the Mirror: A Political Remake of Brunetto Latini's *Li Livres dou Tresor*', in *The Mirror in Medieval and Early Modern Culture: Specular Reflections*, ed. by Nancy M. Frelick (Turnhout: Brepols, 2016), pp. 89–111

Negrón, Luis M. Girón, 'Islamic and Jewish Influences', in *Dante in Context*, ed. by Zygmunt G. Barański and Lino Pertile (Cambridge: Cambridge University Press, 2015), pp. 200–20

Nerva, Mario, 'L'"Imago mundi" di Onorio Augustodunensis e le scienze naturali nel dodicesimo secolo' (unpublished MA thesis, Université de Montréal, 1971)

Neubauer, A., 'Les traductions hébraïques de l'*Image du Monde*', *Romania*, 5 (1876), 129–39

Nichols, Stephen G., 'Introduction: Philology in a Manuscript Culture', *Speculum*, 65 (1990), 1–10

Nunn, George E., 'The *Imago Mundi* and Columbus', *The American Historical Review*, 40 (1935), 646–61

O'Doherty, Marianne, 'Eyewitness Accounts of "the Indies" in the Later Medieval West: Reading, Reception and Re-use (*c.* 1300–1500)' (unpublished PhD thesis, University of Leeds, 2006)

O'Doherty, Marianne, *The Indies and the Medieval West: Thought, Report, Imagination* (Turnhout: Brepols, 2013)

Ó Carragáin, Tomás, *Churches in Early Medieval Ireland: Architecture, Ritual and Memory* (New Haven: Yale University Press, 2010)

O'Loughlin, Thomas, 'An Early Thirteenth-Century Map in Dublin: A Window into the World of Giraldus Cambrensis', *Imago Mundi*, 51 (1999), 24–39

Ommundsen, Åslaug, and Tuomas Heikkilä, eds, *Nordic Latin Manuscript Fragments: The Destruction and Reconstruction of Medieval Books* (London and New York: Routledge, 2017)

Onuma, Yu, 'Convention through Innovation: Marvels in *Topographia Hibernica* by Gerald of Wales', in *Aspetti del meraviglioso nelle letterature medievali: Medioevo latino, romanzo,*

germanico e celtico, ed. by Franca Ela Consolino, Francesco Marzella, and Lucilla Spetia (Turnhout: Brepols, 2016), pp. 69–80
Oschema, Klaus, *Bilder von Europa im Mittelalter*, Mittelalter-Forschungen, 43 (Ostfildern: Thornbecke, 2013)
Oschema, Klaus, *'Europe' in the Middle Ages* (Leeds: ARC Humanities Press, 2023)
Oostrom, Frits van, *Maerlants wereld* (Amsterdam: Prometheus, 1996)
Oostrom, F. D. van, and C. H. J. M. Kneepkens, 'Maerlants *Alexanders geesten* en de *Alexandreïs*: een terreinverkenning', *De niewe taalgids*, 69.6 (1976), 483–500
Pagden, Anthony, 'Europe: Conceptualizing a Continent', in *The Idea of Europe*, ed. by Pagden, pp. 33–54
Pagden, Anthony, ed., *The Idea of Europe: From Antiquity to the European Union* (Cambridge: Cambridge University Press, 2002)
Paniagua, David, '"Soccorri me, che solo non so ire": Solino n aiuto di Fazio degli Uberti', *Cento Pagine*, 3 (2009), 10–19
Park, Hyunhee, *Mapping the Chinese and Islamic Worlds* (Cambridge: Cambridge University Press, 2015)
Parker, Margaret A., 'Juan de Mena's Ovidian Material: an Alfonsine Influence', *Bulletin of Hispanic Studies*, 55 (1978), 5–17
Parker, W. H., 'Europe: How Far?', *The Geographical Journal*, 126 (1960), 278–97
Parkes, Malcom Beckwith, 'The Influence of the Conepts of *Ordinatio* and *Compilatio* on the Development of the Book', in *Medieval Learning and Literature. Essays Presented to Richard William Hunt*, ed. by J. J. G. Alexander and M. T. Gibson, (Oxford: Clarendon Press, 1976), pp. 115–41
Pedrycz, Witold, and Fernando Gomide, *An Introduction to Fuzzy Sets: Analysis and Design* (Cambridge, MA: MIT Press, 1998)
Pelle, Stephen, 'An Old Norse Adaptation of a Christmas Sermon by Honorius Augustodunensis', *Medieval Sermon Studies*, 61 (2017), 44–58
Pellizzari, Achille, *Il Dittamondo e la Divina Commedia. Saggio sulle fonti del 'Dittamondo' e sulla imitazione dantesca nel secolo XIV* (Pisa: Tipografia Editrice Francesco Mariotti, 1905)
Penn, Stephen, 'Chronicon Wirziburgense', in *Encyclopedia of the Medieval Chronicle*, ed. by Graeme Dunphy and Cristian Bratu http://dx.doi.org/10.1163/2213-2139_emc_SIM_00673 (accessed 4 October 2018)
Penrose, Roger, *The Road to Reality: A Complete Guide to the Laws of the Universe* (London: Vintage Books, 2005)
Petrovskaia, Natalia I. 'The Concept of Europe in the Medieval Welsh Geographical Treatise *Delw y Byd*', *Celtic Forum*, 21 (2018), 23–34
Petrovskaia, Natalia I., '*Delw y byd*: une traduction médiévale en gallois d'une encyclopédie latine et la création d'un traité géographique', *Études celtiques*, 39 (2013), 257–77
Petrovskaia, Natalia I., 'Europe and the Holy Land in the British Branch of the *Imago mundi* Tradition', in *Crusading Culture and the Holy Land in Medieval Britain*, ed. by Kathryn Hurlock and Laura Whatley (Turnhout: Brepols, 2022), pp. 41–66
Petrovskaia, Natalia I., 'Explicit and Implicit Multilingualisms. *Imago mundi* and MS Estense α.Q.5.1', in *Multilingual Dynamics of Medieval Literature in Western Europe, c. 1200–c. 1600*, ed. by Bart Besamusca, David Murray, and Lisa Demets (Turnhout: Brepols, forthcoming)
Petrovskaia, Natalia I., 'La formulation tripartite du monde dans les encyclopédies et textes littéraires du Moyen Âge', in *La Formule au Moyen Âge IV / Formulas in Medieval Culture IV*, ed. by E. Louviot, C. Garcia, and S. Morrison, ARTeM Atelier de Recherche sur les Textes Médiévaux, 31 (Turnhout: Brepols, 2022), pp. 197–217

Petrovskaia, Natalia I., 'L'image du monde animalier', in *Mondes animaliers au Moyen Age et à la Renaissance – Tierische Welten im Mittelalter und in der Renaissance – Actes du Colloque international des 8, 9, 10 et 11 mars 2016 à la Maison de la Culture d'Amiens*, ed. by Danielle Buschinger, Florent Gabaude, Marie-Geneviève Grossel, Jürgen Kühnel, and Mathieu Olivier (Amiens: Presses du 'Centre d'Etudes Médiévales de Picardie', 2016), pp. 313–26

Petrovskaia, Natalia I., 'The Limits of Medieval Translation: Juan de Mena's Use of the *Imago mundi* in the *Laberinto de Fortuna*', *Translation Matters*, 5 (2023), 19–30, DOI: 10.21747/21844585/tm5_2a2

Petrovskaia, Natalia I., *Medieval Welsh Perceptions of the Orient* (Turnhout: Brepols, 2015)

Petrovskaia, Natalia I., *Modern Afterlives of Old Irish Travel Narratives: From Gulliver to Star Trek* (Leiden: Brill, forthcoming).

Petrovskaia, Natalia I., 'Mythologizing the Conceptual Landscape: Religion and History in *Imago Mundi, Image du Monde* and *Delw y Byd*', in *Landscape and Myth in North-Western Europe*, ed. by Matthias Egeler, Medieval Texts and Cultures of Northern Europe (Turnhout: Brepols, 2019), pp. 195–211

Petrovskaia, Natalia I., 'Re-Writing Parts of Europe in Some Vernacular Adaptations of the *Imago Mundi*', in *Medieval Translatio: Interdisciplinary Studies in the Translation and Transfer of Language, Culture, Literature*, ed. by Massimiliano Bampi and Stefanie Gropper, Modes of Modification, 2 (Brelin: DeGruyter, 2024), pp. 81–101

Petrovskaia, Natalia I., '*Translatio* and Translation: The Duality of the Concept from the Middle Ages to the Early Modern Period', *Doshisha Studies in English*, 99 (2018), 119–36

Petrovskaia, Natalia I., 'Translation and Transmission of Texts in Medieval Europe: Two Aspects of *Translatio*', in *Literature, Science and Religion: Textual Transmission in Medieval and Early Modern Europe*, ed. by Manuel Bellmunt Serrano and Joan Mahiques Climent, Problemata Litteraria, 88 (Kassel: Edition Reichenberger, 2020), pp. 375–90

Petrovskaia, Natalia I., 'Which "Pagans"?: The Influence of the Crusades on Battle Narratives in Britain, Ireland and Scandinavia', in *Writing Battles: New Perspectives on Warfare and Memory in Medieval Europe*, ed. by Máire Ní Mhaonaigh, Rory Naismith, and Elizabeth Ashman Rowe (London: Bloomsbury Academic, 2020), pp. 147–64

Petrovskaia, Natalia I., '地図を読む：ヨーロッパの世界地図、中世からルネサンスにかけて' (= 'Reading Maps: European World Maps, from the Middle Ages to the Renaissance'), *書物学 Bibliology*, 13 (2018), 1–6

Picone, M., ed., *L'encyclopedismo medievale* (Ravenna: Longo, 1994)

Plate, L., and H. G. Els Rose, 'Rewriting, a Literary Concept for the Study of Cultural Memory: Towards a Transhistorical Approach to Cultural Remembrance', *Neophilologus*, 97 (2013), 611–625

Pocock, J. G. A., *The Discovery of Islands: Essays in British History* (Cambridge: Cambridge University Press, 2005)

Poppe, Erich, and R. Reck, 'A French Romance in Wales: *Ystorya Bown o Hamtwn*: Processes of Medieval Translations', *Zeitschrift für celtische Philologie*, 55 (2006), 122–80; 56 (2008), 129–64

Post, Chandler Rathfon, 'The Sources of Juan de Mena', *Romanic Review*, 3 (1912), 223–79

Pratt, Karen, 'Medieval Attitudes to Translation and Adaptation: The Rhetorical Theory and the Poetic Practice', in *The Medieval Translator II*, ed. by Roger Ellis (London: Queen Mary University of London, 1991), pp. 1–27

Puppo, Dario del, 'Fazio degli Uberti', in *Medieval Italy: An Encyclopedia*, ed. by Christopher Kleinhenz (New York and London: Routledge, 2004), pp. 334–35

Quinn, Judy, 'From Orality to Literary in Medieval Iceland', in *Old Icelandic Literature and Society*, ed. by Margaret Clunies Ross (Cambridge: Cambridge University Press, 2000), pp. 30–60

Rabaté, Jean-Michel, ed., *The Cambridge Companion to Lacan* (Cambridge: Cambridge University Press, 2003)

Rachetta, Maria Teresa, 'Brunetto Latini, La Storia Universale e la letteratura francese di matrice erudita del primo xiii secolo', in *Dante e la cultura fiorentina. Bono Giamboni, Brunetto Latini e la formazione intellettuale deil laici*, ed. by Zygmunt G. Barański, Theodore J. Chachey, Jr., and Luca Lombardo (Rome: Salerno, 2019), pp. 101–132

Relaño, Francesc, *The Shaping of Africa: Cosmographic Discourse and Cartographic Science in Late Medieval and Early Modern Europe* (Aldershot: Ashgate, 2002)

Raymo, Robert R., and Elaine E. Whitaker, eds, *The Mirroure of the Worlde: A Middle English Translation of Le miroir du monde*, Medieval Academy Books, 106 (Toronto: University of Toronto Press for the Medieval Academy of America, 2003)

Reynolds, Roger E., 'Further Evidence for the Irish Origin of Honorius Augustodunensis', *Vivarium*, 7.1 (1969), 1–7

Ribémont, Bernard, 'Brunetto Latini, le *Livre dou Tresor* et l'histoire sainte: une réception du *De ortu et obitu patrum* d'Isidore de Séville', *Cahiers de Recherches Médiévales*, 16 (2008), 135–58

Ribémont, Bernard, 'L'encyclopedisme médiéval: de la définition d'un genre à son apogée. Sur la pertinence des notions d'apogée et de décadence', in *De Natura Rerum*, ed. by Ribémont, pp. 11–68

Ribémont, Bernard, 'L'inconnu géographique des encyclopédies médiévales', *Cahiers de recherches médiévales*, 3 (1997), 101–11

Ribémont, Bernard, *De Natura Rerum : Etudes sur les encyclopédies médiévales* (Orleans: Paradigme, 1995)

Ribémont, Bernard, *Les origines des encyclopédies médiévales: D'Isidore de Seville aux Carolingiens* (Paris: Champion, 2001)

Ribémont, Bernard, *La 'Renaissance' du XIIe siècle et l'encyclopédisme*, Essais sur le Moyen Âge, 27 (Paris, 2002)

Ribémont, Bernard, 'Statut de l'astronomie et evolution des connaissances sur le cosmos chez les vulgarizateurs médiévaux: le cas de quelques encyclopédies en langue vernaculaire', in *Observer, lire, écrire le ciel au Moyen Âge. Actes du colloque d'Orléans, 22–23 avril 1989*, ed. by Bernard Ribémont (Paris, 1991), pp. 283–300

Ribémont, Bernard, 'On the Definition of An Encyclopaedic Genre in the Middle Ages', in *Pre-Modern Encyclopaedic Texts*, ed. by Binkley, pp. 47–61

Richardson, John, 'Provincial Administration', in *The Oxford Handbook of Roman Law and Society*, ed. by Paul J. du Plessis, Clifford Ando, and Kaius Tuori (Oxford: Oxford University Press, 2016), pp. 111–23

Rigall, Juan Casas, 'Razas humanas portentosas en las partidas remotas del mundo (del Benjamín de Tutela a Cristóbal Colón', in *Maravillas, peregrinaciones y utopías: Literatura de viajes en el mundo románico*, ed. by Rafael Beltrán (Valencia, 2002), pp. 253–90

Riley-Smith, Jonathan S. C., *The Crusades: A Short History* (London: Athlone, 1987)

Rios, José Amador de los, *Historia Critica de Literatura Española*, 7 vols (Madrid: José Rodriguez, 1863)

Röder, Gabriele 'Die chansons de geste in der altnordischen *Karlamagnús saga*: Übersetzungen oder Adaptationen', in *The Medieval Translator: Traduire au moyen âge*, ed. by Roger Ellis, René Tixier, and Bernd Weitemeier, The Medieval Translator, 6 (Turnhout: Brepols, 1998), pp. 134–58

Rösli, Lukas, 'Spatial Studies', in *Handbook of Pre-Modern Nordic Memory Studies: Interdisciplinary Approaches*, ed. by Jürg Glauser, Pernille Hermann, and Stephen A. Mitchell, 2 vols (Berlin: De Gruyter, 2018), I, pp. 274–83

Romm, James S., 'Continents, Climates, and Cultures: Greek Theories of Global Structure', in *Geography and Ethnography: Perceptions of the World in Pre-Modern Societies*, ed. by Kurt A. Raaflaub and Richard J. A. Talbert (Chichester: Wiley–Blackwell, 2010), pp. 215–35

Romm, James S., *The Edges of the Earth in Ancient Thought: Geography, Exploration, and Fiction* (Princeton, NJ: Princeton University Press, 1992)

Ruberg, Uwe, '*Mappae mundi* des Mittelalters im Zusammenwirken von Text und Bild', in *Text und Bild: Aspekte des Zusammenwirkens zweier Künste in Mittelalter und früher Neuzeit*, ed. by Christel Meier and Uwe Ruberg (Wiesbaden: Ludwig Reichert, 1980), pp. 550–92

Ruhe, Ernstpeter, *Elucidarium und Lucidaires: zur Rezeption des Werks von Honorius Augustodunensis in der Romania und in England*, Wissensliteratur im Mittelalter, 7 (Wiesbaden, Reichert: 1993)

Ruiz, Teofilo F., *Spanish Society 1400–1600* (London and New York: Routledge, 2014)

Sacchi, Luca, 'L'edizione di testi modulari: il caso del *Lucidario* di Sancho IV di Castiglia', in *Prassi ecdotiche: esperienze editoriali su testi manoscritti e testi a stampa*, ed by A. Cadioli and Paolo Chiesa (Milan: Cisalpino, 2008), pp. 155–70

Sacerdote, Gustavo, 'The Ninth Mehabbereth of Emanuele da Roma and the *Tresor* of Piere de Corbiac', *Jewish Quarterly Review*, 7 (1895), 711–28

Saenger, Paul, *Space Between Words: The Origins of Silent Reading* (Stanford, CA: Stanford University Press, 1997)

Sáenz-López Pérez, Sandra, 'El "otro" en la cartografía bajomedieval: aportaciones desde la lectura de los mapas', in *Através do olhar do Outro: Reflexões acerca de sociedade europeia (séculos XII–XV)*, ed. by José Albuquerque Carreiras, Giulia Rossi Vairo, and Kristjan Toomaspoeg (Tomar: Instituto Politécnico de Tomar, 2018), pp. 135–56

Said, Edward, *Orientalism* (New York: Pantheon, 1978)

Sanford, Eva Mathews, 'Honorius *Presbyter* and *Scholasticus*', *Speculum*, 23 (1948), 397–425

Sarton, George, *Introduction to the History of Science, Volume II: From Rabbi Ben Ezra to Roger Bacon* (Baltimore: Williams and Wilkins for the Carnegie Institution of Washington, 1931)

Sayle, C. E., *Early English Printed Books in the University Library Cambridge (1475 to 1640)*, 4 vols (Cambridge: Cambridge University Press, 1900–7), I (1900)

Schipperges, H., 'Honorius und die Naturkunde des 12. Jahrhunderts', *Suddhovs Archiv. Zeitschrift für Wissenschaftsgeschichte*, 42 (1958), 71–82

Schlauch, Margaret, *Romance in Iceland*, 2nd edn (New York, 1973)

Schmale, Wolfgang, 'Europe: Eighteenth-Century Definitions', in *Bordering Early Modern Europe*, ed. by Maria Baramova, Grigor Boykov, and Ivan Parvev (Wiesbaden: Harrazzowitz Verlag, 2015), pp. 79–93

Schmidt, Gleb, 'From Manual to Best-Seller: The History of Honorius Augustodunensis's *Elucidarium*', in *Books of Knowledge in Late Medieval Europe: Circulation and Reception of Popular Texts*, ed. by Pavlína Cermanová and Václav Žůrek, Utrecht Studies in Medieval Literacy, 52 (Turnhout: Brepols, 2021), pp. 137–63

Schneidmüller, Bernd, 'Die mittelalterichen Destillationen Europas aus der Welt', in *Europa in der Welt des Mittelalters: Ein Colloquium für und mit Michael Borgolte*, ed. by Tillman Lohse and Benjamin Scheller (Berlin; De Gruyter, 2014), pp. 11–32

Schneidmüller, Bernd, 'Die mittelalterlichen Konstruktionen Europas: Konvergenz und Differenz', in *'Europäische Geschichte' als historiographisches Problem*, ed. by H. Ducchardt and A. Kunz, Veröffentlichungen des Instituts für Europäische Geschichte, Abteilung Universalgeschichte Beiheft 42, (Mainz, 1997), pp. 5–24

Schneidmüller, Bernd, 'Europa im Mittelalter. Vorstellungen und Forschungsaufgaben', in *Integration und Transformation in Europa. Beiträge aus dem Forschungsschwerpunkt 'Integration und*

Transformation in Europa (ITE)', Forschungsforum. Berichte aus der Otto-Friedrich-Universität Bamberg, 9 (Bamberg, 1999), pp. 6–16

Schorbach, Karl, *Studien über das deutsche Volksbuch Lucidarius und seine Bearbeitungen in fremden Sprachen*, Quellen und Forschungen zur Sprach- und Kulturgeschichte der germanischen Völker, 74 (Strassburg: Karl J. Trübner, 1894)

Schöller, Bettina, 'Transfer of Knowledge: *Mappae Mundi* Between Texts and Images', *Peregrinations: Journal of Medieval Art and Architecture*, 4 (2013), 42–55

Seniff, Dennis P., 'Orality and Textuality in Medieval Castilian Prose', *Oral Tradition*, 2 (1987), 150–71

Serchuk, Camille, 'Gaul Undivided: Cartography, Geography and Identity in France', in *Mapping Medieval Geographies: Geographical Encounters in the Latin West and Beyond*, ed. by Keith Lilley (Cambridge: Cambridge University Press, 2013), pp. 175–200

Shohat, Ella, and Robert Stam, *Race in Translation: Culture Wars around the Postcolonial Atlantic* (New York and London: NYU Press, 2012)

Sif Ríkharðsdóttir, 'The Norse Roland in Context', in *Charlemagne in the Norse and Celtic Worlds*, ed. by Helen Fulton and Sif Ríkharðsdóttir (Cambridge: D. S. Brewer, 2022), pp. 50–65

Silvi, Christine, 'La mise en recueil de l'*Image du monde* (rédaction en prose) dans les premiers imprimés: diffusion et réception d'une *livre de clergie* médiéval aux XVe et XVIe siècles', *Seizième siècle*, 13 (2017), 293–28

Silvi, Christine, 'Petite histoire d'une appropriation réussie: François Buffereau plagiaire de Gossuin de Metz', *Revue belge de Philologie et d'Histoire*, 94 (2016), 661–89

Silvi, Christine, 'Les variations sur le nom dans la *Mappemonde* de maître Gossuin de Metz', in *Par les mots et les textes. Mélanges de langue, de littérature et d'histoire des sciences médiévales offerts à Claude Thomasset*, ed. by Danielle Jacquart, Danièle James-Raoul and Olivier Soutet (Paris, 2005), pp. 679–91

Simek, Rudolf, *Altnordische Kosmographie: Studien und Quellen zu Weltbild und Weltbeschreibung in Norwegen und Island vom 12. bis zum 14. Jahrhundert* (Berlin and Boston: De Gruyter, 2012)

Simek, Rudolf, *Heaven and Earth in the Middle Ages: The Physical World Before Columbus*, trans. by Angela Hall (Woodbridge: Boydell, 1996)

Simek, Rudolf, 'Die Quellen der *Eiríks saga víðförla*', *Skandinavistik: Zeitschrift für Sprache, Literatur und Kultur der Nordischen Länder*, 14 (1984), 109–114

Singerman, Robert, *Jewish Translation History: A Bibliography of Bibliographies and Studies* (Amsterdam: John Benjamins, 2002)

Spiegel, Gabrielle M., 'History, Historicism, and the Social Logic of the Text in the Middle Ages', *Speculum*, 65 (1990), 59–86

Spinosa, Giacinta, '*Translatio studiorum* through Philosophical Terminology', in *Translatio studiorum: Ancient, Medieval and Modern Bearers of Intellectual History*, ed. by Marco Sgabi (Leiden and New York: Brill, 2012), pp. 73–89

Spivak, Gayatry Chakravorty, 'Can the Subaltern Speak?', in *Marxism and the Interpretation of Culture*, ed. by Cary Nelson and Lawrence Grossberg (Urbana and Chicago: University of Illinois Press, 1988), pp. 271–313

Strayer, Joseph R., 'France: The Holy Land, the Chosen People, and the Most Christian King', in *Medieval Statecraft and the Perspectives of History: Essays by Joseph R. Strayer* (Princeton, NJ: Princeton University Press, 1971) pp. 300–14

Street, Florence, 'La vida de Juan de Mena', *Bulletin Hispanique*, 55 (1953), 149–73

Strzelcsyk, Jerzy, 'Gervase of Tilbury (*c.* 1165–*c.* 1234)', in *Trade, Travel, and Exploration in the Middle Ages: An Encyclopedia*, ed. by John Block Friedman and Kristen Mossler Figg, with Scott D. Westrem and Gregory G. Guzman (New York and London: Routledge, 2000), pp. 227–28

Stückelberger, Alfred, 'Das Europabild bei Ptolemaios', in *Europa im Weltbild des Mittelalters. Kartographische Konzepte*, ed. By Ingrid Baumgärtner and Hartmut Kugler, Orbis mediaevalis, 10 (Berlin: Akademie Verlad, 2008), pp. 31–44

Stuckey, Jace, 'Introduction', in *The Eastern Mediterranean Frontier of Latin Christendom*, ed. by Jace Stuckey, The Expansion of Latin Europe, 1000–1500, 6 (Farnham and Burlington: Ashgate, 2014), pp. xvii–xxxv

Sturlese, Loris, 'Zwischen Anselm und Johannes Scotus Eriugena. Der seltsame Fall des Honorius, des Mönchs von Regensburg', *Historia philosophiae medii aevi: Studien zur Geschichte der Philosophie des Mittelalters*, ed. by B. Mojisch and O. Pluta (Amsterdam: John Benjamins, 1991), pp. 927–51

Sundby, Thor, *Della vita e delle opere di Brunetto Latini*, trans. by R. Renier (Florence: Successori Le Monnier, 1884)

Sverrir Jakobsson, 'On the Road to Paradise: "Austrvegr" in the Icelandic Imagination', in *The Fantastic in Old Norse/Icelandic Literature: Sagas and the British Isles. Preprint papers of the 13th international Saga Conference, Durham and York, 6th–12th August, 2006*, ed. by John McKinnell, David Ashurst, and Donata Kick (Durham: Durham University Press, 2006), pp. 935–43

Svanhildur Óskarsdóttir, 'Prose of Christian Instruction', in *A Companion to Old Norse-Icelandic Literature and Culture*, ed. by Rory McTurk, Blackwell Companions to Literature and Culture, 31 (Malden, MA, Oxford and Victoria, 2005) pp. 338–53

Svanhildur Óskarsdóttir, 'Universal History in the Fourteenth-Century Iceland: Studies in AM 764 4to' (unpublished PhD thesis, University of London, 2000)

Talbert, Richard J. A., and Richard W. Unger, eds, *Cartography in Antiquity and the Middle Ages: Fresh Perspectives, New Methods*, Terminology and Change in History, 10 (Leiden and Boston, 2008)

Talmor, Ezra, 'Reflections on the Rise and Development of the Idea of Europe', *History of European Ideas*, 1 (1980), 63–66

Termini, Settimo, 'On Some "Family Resemblances" of Fuzzy Set Theory and Human Sciences', in *Soft Computing in Humanities and Social Sciences*, ed. by R. Seising and V. Sanz González, Studies in Fuzziness and Soft Computing, 273 (Berlin and Heidelberg: Springer, 2012), pp. 39–54

Tita, Silvia, 'Brunetto Latini's *Li livres dou Tresor*: The Translation of a Political System', *Linguistic and Philosophical Investigations*, 13 (2014), 288–96

Toynbee, Paget, 'Brunetto Latino's obligatons to Solinus', *Romania*, 89 (1894), 62–77

Trevor-Roper, Hugh, 'Fernand Braudel, the Annales, and the Mediterranean', *The Journal of Modern History*, 4 (1972), 468–79

Tristram, Hildegard, ed., *Sex aetates mundi: die Weltzeitalter bei den Angelsachsen und den Iren. Untersuchungen und Texte*, Anglistische Forschungen, 165 (Heidelber, 1985)

Tuan, Yi-Fu, *Space and Place: The Perspective of Experience* (Minneapolis and London: University of Minnesota Press, 1977)

Turville-Petre, G., *Origins of Icelandic Literature* (Oxford: Clarendon Press, 1953)

Twomey, Michael W., 'Honorius Augustodunensis', *Trade, Travel, and Exploration in the Middle Ages: An Encyclopedia*, ed. by John Block Friedman and Kristen Mossler Frigg, with Scott D. Westrem and Gregory G. Guzman (New York and London: Routlledge, 2000), pp. 259–61

Twomey, Michael W., 'Honorius Augustodunensis (c. 1080–c. 1140)', *Routledge Resources Online: Medieval Studies* (Routledge, 2022), DOI: 10.4324/9780415791182-RMEO96-1

Twomey, Michael W., 'Inventing the Encyclopedia', in *Schooling and Society: The Ordering and Reordering of Knowledge in the Western Middle Ages*, ed. by Alasdair A. MacDonald and Michael W. Twomey, Groningen Studies in Cultural Exchange, 6 (Leuven, Paris, and Dudley, MA: Peeters, 2004), pp. 73–92

Twomey, Michael W., 'Medieval Encyclopedias', in *Medieval Christian Literary Imagery. A Guide to Interpretation*, ed. by R. E. Kaske (University of Toronto Press, 1988), pp. 182–215

Uebel, Michael, ed., *Ecstatic Transformation: On The Uses of Alterity in the Middle Ages*, The New Middle Ages (New York: Palgrave Macmillan, 2005)

Van Duzer, Chet, 'Cartographic Invention: The Southern Continent on Vatican MS Urb. Lat. 274, Folios 73v–74r (*c.* 1530)', *Imago mundi*, 59 (2007), 193–222

Van Duzer, Chet, 'A Neglected Type of Medieval Mappamundi and Its Re-imaging in the Mare Historiarum (BnF MS Lat. 4915, Fol. 26v)', *Viator*, 43 (2012), 277–301

Van Duzer, Chet, and Sandra Sáenz-López Pérez, '*Tres filii Noe diviserunt orbem post diluvium*: The World Map in British Library Add. MS 37049', *Word & Image*, 26 (2009), 21–39

Voigts, Linda Ehrsam, 'Scientific and Medical Books', in *Book Production and Publishing in Britain 1375–1475*, ed. by Jeremy Griffiths and Derek Pearsall (Cambridge: Cambridge University Press, 1989), pp. 345–402

Von Kelle, J., 'Untersuchungen über den nicht nachweisbaren Honorius Augustodunensis presbyter und scholasticus und die ihm zugeschiebenen Werke', in *Sitzungsberichte des philosophisch-historischen Klasse der kaiserlichen Akademie der Wissenschaften zu Wien*, 102 (1905), fasc. 2

Walker, Roger M., *Tradition and Technique in 'El Libro del Cavallero Zifar'* (London, 1974)

Wallace, David J., 'Brunetto Latini', in *Medieval France. An Encyclopedia*, ed. by William W. Kibler, Grover A. Zinn, Lawrence Earp, and John Bell Henneman, Jr., Routledge Revivals (Abingdon: Routledge, 2016), pp. 151–52

Wallace, William, 'Where Does Europe End? Dilemmas of Inclusion and Inclusion', in Jan Zielonka ed., *Europe Unbound: Enlarging and Reshaping the Boundaries of the European Union* (New York: Routledge, 2002)

Waters, Claire M., *Translating Clergie: Status, Education, and Salvation in Thirteenth-Century Vernacular Texts* (Philadelphia: University of Pennsylvania Press, 2016)

Watts, Pauline Moffitt 'Prophecy and Discovery: On the Spiritual Origins of Christopher Columbus's "Enterprise of the Indies"', *The American Historical Review*, 90 (1985), 73–102

Weeda, Claire V., *Ethnicity in Medieval Europe, 950–1250: Medicine, Power and Religion* (York: York Medieval Press, 2021)

Weeda, Claire V., 'Images of Ethnicity in Later Medieval Europe' (unpublished PhD dissertation, Universiteit van Amsterdam, 2012)

Weiss, Julian, 'Remembering Spain in the Medieval European Epic: A Prospect', in *Locating the Middle Ages: The Spaces and Places of Medieval Culture*, ed. by Julian Weiss and Sarah Salih (London: King's College London Medieval Studies, 2012), pp. 67–82

Wesseling, H., *A Cape of Asia: Essays on European History* (Leiden: Leiden University Press, 2011)

Westrem, Scott D., *Broader Horizons: A Study of Johannes Witte de Hese's Itinerarius and Medieval Travel Narratives*, Medieval Academy Books, 105 (Cambridge, MA: The Medieval Academy, 2001)

Wetzstein, Thomas, 'New Masters of Space: The Creation of Communication Networks in the West (Eleventh-Twelfth Centuries)', in *Space in the Medieval West*, ed. by Cohen and Madeline, pp. 115–32

Whalen, Brett Edward, *The Two Powers: The Papacy, the Empire, and the Struggle for Sovereignty in the Thirteenth Century* (Philadelphia: University of Pennsylvania Press, 2019),

Wiedemann, Benedict, 'The Joy of Lists: The *Provinciale Romanum*, Tribute and *Ad limina* Visitation to Rome', *Revue d'histoire ecclésiastique*, 115 (2020), 61–97

Wick, Alexis, *The Red Sea: In Search of Lost Space* (Oakland: University of California Press, 2016)

Wintle, Michael, 'The Early Modern Iconography of Europe: Visual Images and European Identity', in *Contesting Europe: Comparative Perspectives on Early Modern Discourses on*

Europe, 1400–1800, ed. by Nicolas Detering, Clementina Marsico, and Isabella Walser-Bürgler (Leiden and Boston: Brill, 2020), pp. 54–76

Wintle, Michael, *The Image of Europe: Visualising Europe in Cartography and Iconography Throughout the Ages* (Cambridge: Cambridge University Press, 2009)

Wintle, Michael, 'Renaissance Maps and the Construction of the Idea of Europe', *Journal of Historical Geography*, 25 (1999), 137–65

Wojciechowski, Leszek, 'Geographical Descriptions of Spain in Latin Medieval Encyclopedias: From Isidore of Seville to Vincent of Beauvais (VII-XIII C.)', *Roczniki Humanistyczne*, 67 (2019), 31–55, DOI: 10.18290/rh.2019.67.2-3

Wolf, Armin, 'The Ebstorf *Mappamundi* and Gervase of Tilbury: The Controversy Revisited', *Imago mundi*, 64 (2012), 1–27

Woodward, David, 'Geography', in, ed., *The Cambridge History of Science. Volume 2. Medieval Science* David C. Lindberg and Michael H. Shank (Cambridge: Cambridge University Press, 2013), pp. 548–68

Woodward, David, 'Medieval *Mappaemundi*', in *History of Cartography I*, ed. by Harley and Woodward, pp. 286–370

Woodward, David, 'Reality, Symbolism, Time and Space in Medieval World Maps', *Annals of the Association of American Geographers*, 75 (1985), 510–21

Yeager, Suzanne M., 'The World Translated: Marco Polo's *Le Devisement dou monde, The Book of Sir John Mandeville*, and Their Medieval Audiences', in *Marco Polo and the Encounter of East and West*, ed. by Suzanne Conklin Akbari and Amilcare Iannucci, with the assistance of John Tulk (Toronto, Buffalo and London: University of Toronto Press, 2008), pp. 156–81

Yearwood, Peter J., 'Continents and Consequences: The History of a Concept', *Journal of Global History*, 9 (2014), 329–56

Zadeh, Lotfi A., 'Fuzzy Sets', *Information and Control*, 8 (1965), 338–53

Zambarelli, Luigi, *L'imitazione dantesca nel 'Dittamondo' e nelle liriche morali di Fazio degli Uberti* (Rome: Atena, 1942)

Zhu, Lifeng, Weiwei Xia, Jia Liu, and Aiguo Song, 'Visualizing Fuzzy Sets Using Opacity-Varying Freeform Diagrams', *Information Visualization*, 17 (2017), 146–60

Zonta, Mauro, 'Medieval Hebrew Translations of Philosophical and Scientific Texts: *A Chronological Table*', in *Science in Medieval Jewish Cultures*, ed. by Gad Freudenthal (Cambridge: Cambridge University Press, 2012), pp. 17–73

Zucker, Arnaud, 'Introduction', in *Encyclopédire: Formes de L'Ambition Encyclopédique dans l'Antiquité et au Moyen Âge*, ed. by Arnaud Zucker (Turnhout: Brepols, 2013), pp. 11–32

Electronic Resources

ARLIMA: Archives de littérature du moyen âge https://www.arlima.net/eh/honorius_augustodunensis (accessed 11 April 2024)

British Library Incunabula Short Title Catalogue http://www.bl.uk/catalogues/istc/ (accessed 20 September 2018)

Charles M. Schulz Museum Online Collections Database https://schulzmuseum.pastperfectonline.com/ (accessed 24 July 2020)

Dictionnaire du Moyen Français (1330–1500) http://www.atilf.fr/dmf/ (accessed 26 April 2020)

The Digital Bodleian www.digital.bodleian.ox.ac.uk/ (accessed 11 April 2024)

The Digital Walters http://www.thedigitalwalters.org/ (accessed 11 April 2024)

EEBO – Early English Books Online https://eebo.chadwyck.com/home (accessed 11 April 2024)
Electronic Dictionary of the Irish Language, based on *Dictionary of the Irish Language*, ed by E. G. Quin et al. (Dublin: Royal Irish Academy, 1913-1976) https://dil.ie/ (accessed 2 November 2024).
Gallica (BnF) https://gallica.bnf.fr (accessed 4 October 2023)
Geiriadur Pryfisgol Cymru: A Dictionary of the Welsh Language, Cardiff 1950–2002 https://geiriadur.ac.uk/gpc/gpc.html (accessed 4 July 2024)
Handrit.is National and University Library of Iceland 2009–2021 (Norse manuscripts from the manuscript department of the National Library – University Library, the Árni Magnússon Institute for Icelandic Studies and the Arnas Museum in Copenhagen) https://handrit.is/is/ (accessed 3 April 2023)
Handschriftenportal https://handschriftenportal.de/ (Staatsbibliothek zu Berlin – Preußischer Kulturbesitz, Universitätsbibliothek Leipzig, Herzog August Bibliothek Wolfenbüttel and Bayerischer Staatsbibliothek München) (accessed 4 July 2024)
Images of the World: Manuscripts Database of the Imago Mundi tradition, https://imagomundi.hum.uu.nl/ by Natalia I. Petrovskaia and Kiki Calis (published online in 2019; accessed 28 March 2024)
Österreichische Nationalbibliliothek online catalogue https://search.onb.ac.at/primo-explore/search?vid=ONB (accessed 28 March 2024)
Parker Library on the Web: Manuscripts in the Historic Parker Library at Corpus Christi College, Cambridge https://parker.stanford.edu/parker/catalog/jb848tp9919 (accessed 9 April 2024)
Perseus Digital Library, ed. by Gregory R. Crane. Tufts University http://www.perseus.tufts.edu (accessed 11 April 2024)
Repertorium. 'Geschichtsquellen des deutschen Mittelalters' (Bayerische Akademie der Wissenschaften) https://www.geschichtsquellen.de/repOpus_02846.html (accessed 11 April 2024)
Routledge Resources Online – Medieval Studies https://www.taylorfrancis.com/search?context=rroms (accessed 24 February 2023)

Index

Abydos 124
adaptations 57, 58–59, 63, 68, 155, 168, 171
of *Elucidarium*, *see Elucidarium*
of *Imago mundi* 13, 30, 32, 33, 51, 53–6, 155–59, 164, 168, 176
Latin 53, 77–79, 99, 159, 163,
vernacular 30, 32, 33, 42, 45, 53, 54, 57, 59, 61, 63, 65, 67–69, 74–77, 94, 96, 99, 105, 123, 129–30, 135, 136, 138, 139, 149, 150, 151, 153, 157–61, 163, 167
see also *Imago mundi*; rewriting; translations
Aeneas 47
Aeolian Islands 124
Africa 23, 25, 32, 37, 45–48, 69, 88n20, 90, 102, 103, 105, 106, 109–16, 118, 122, 123–26, 130, 136n22, 137–40, 142–43, 162, 165n29
air 37, 45, 49, 74
see also four elements
Alania 89, 90, 110
Albania 70n77, 173n6
Albano 100
Alfonso X, king of Spain 64, 65, 133, 163
Alps 92, 107, 113, 145–47, 149, 151
Amazons 73, 173n6
Americas 166, 171
Amsterdam 71
anachronism 28, 38, 94, 96, 103, 139n34, 160
see also archaism
Andalucía 98
Anglia *see* England
Anselm of Canterbury 40, 42, 64, 168
Aonia 90
Appenines 147, 149
Aquitaine 149
Aragon 74
Armenia 70n77, 173n6
archaism 17, 33, 44, 83, 87–96, 98, 101n70, 104, 155, 159, 160, 170
Ariosto, Ludovico 13–14, 18, 32, 33, 141
Orlando Furioso 13–14, 18, 39, 53, 83, 127, 141, 143
Aristotle 27
Asia 23, 25, 29, 37, 45–48, 69n70, 88n20, 90, 102, 103, 106, 108, 109, 111–12, 113n31, 116, 118, 119n56, 124–27, 130–31, 137–38, 142–43, 162, 165n29
See also Europe, boundaries of
Asia Minor 90, 130
astronomy 37, 45, 50
Atlantis 125
auctoritas 58n23, 104
see also authority
audience *see* readers
Augustine 23, 45n30
aurality 129, 131–35
authority 23, 40, 55–58, 60, 64, 78, 83–85, 87, 103–04, 141, 164, 170
see also auctoritas
authorship *see Imago mundi*
Autun 40, 41
Azov (sea) 95–97

Babylon 73
Bactria 173n6
Balearic Islands 125
Bartholomaeus Anglicus 16n11, 39n6, 57, 59, 73n87, 136
De proprietatibus rerum 39n6, 59, 73n87, 136
Bavaria 44, 91–93
Bede, the Venerable 40, 41n11, 50n49, 51, 64, 88
Belcalzèr, Vivaldo 59, 73n87, 136, 159, 163
Betica 98
Bible 47n33, 50, 69, 73, 86, 114, 116, 119
Black Sea 96–97
Boiardo, Matteo Maria 53, 141
Orlando Innamorato 53, 141
Book of Leccan 95n47
borders 13–14, 17, 18, 22, 24, 25, 28–33, 39, 48–49, 61, 91–93, 96, 99, 104, 105–13, 115, 120–22, 124n76, 125–28, 129, 142, 150, 152–53, 155, 160–64, 167, 172
see also Europe, boundaries of
Brendan, saint 40n9, 125
Brindisi 146, 148
Britain 49n41, 99, 118–19, 123–25, 136, 143, 151n66, 156, 174
Bruges 70
Brunetto Latini 30n58, 53, 57, 59, 71–73, 75, 83, 99–103, 105, 108, 109, 114, 124n76, 127, 129, 136, 141, 151–53, 159, 160, 163, 171n49
Livre dou Trésor 30n58, 53, 55, 59, 71–74, 75, 100–04, 105, 108, 109, 121n61, 124n76, 127, 136, 141, 151–52, 159, 163, 171n49
Tesoretto 60n31, 72n83, 75, 141
Bythinia 90, 103, 173 nn. 6–8
Byzantium 109n13, 116

Caffa *see* Feodosia
Calabria 151
Campania 146–48, 151
Canterbury 40, 42, 168
Capadocia 173n6
Capua 146–47
Carinthia 92
Cartagena 98
Carthage 47, 146, 148

cartography 14–15, 22, 24–30, 37, 47–48, 58, 86n11, 111, 120, 150, 152, 156, 167
see also mappae mundi
Carolingian 88
Cassiodorus 50n49, 88
Castilian 63, 74, 75, 123, 131–33, 163
Catalan Atlas *see* Ecumene Chart
Catalonia 98
Caxton, William 58, 59, 69n68, 70, 71, 72n84, 84n3, 112, 118, 120–21, 157, 159, 164, 167, 170, 171
The Mirrour of the World 70, 112–13, 159, 169–70
Charlemagne 134–35
Chios 124n78
Christ 116
Christendom 20–21, 101, 104, 113n32, 114, 115n37, 160
Christianity 31, 50, 73, 100–03, 113, 114
Christianus 44, 173
Chronicon Wirziburgense 50n49
chronology *see* time
Church *see* Christianity
churches 100
chronotope 13, 24–28, 133, 164, 165
see also space-time
Cidon 124
Clement I, pope 173n8
clergie 57
see also livre de clergie
Corsica 48, 125, 151
Colubria 125
Columbus, Christopher 60, 166, 167, 170
compilatio 57, 72, 78, 143
Cos 124
cosmography 23, 67n58, 72n83, 86n11
Creation 45, 50, 73, 150
Cresques, Elisha ben Abraham 58, 63n35, 74
Crete 124
Crimean Peninsula 96–97
Cyclades 124
Cyprus 124
Cyrnus 125
Cythera 124

Dacia 89, 90, 110
Dalmacia 90
Dania *see* Denmark
Dante Alighieri 32, 60, 75, 140, 142 n.42
Danube (river) 92, 110
Delos 124
Delw y Byd 42n16, 43, 44n24, 53, 58, 60, 64–65, 77, 96–98, 109n16, 132, 144–45, 159–60
Denmark 85
deviser 106–08, 127
Dinus saga drambláta 86, 87n15
division 105, 107, 109n16, 111
See also deviser; world, structure of
Divisiones mundi see Perot de Garbelei
Don (river) 93, 95, 106, 110, 121, 122

earth 22, 37, 45, 46, 48, 51, 74, 107, 108, 119,
see also four elements
Earthly Paradise 76, 86
east 47n37, 76, 89, 100, 115, 118, 139, 140, 143, 151
Ebosus 125
Ecumene Chart 58, 59, 63n35, 74, 159
Einstein, Albert 27
Eiríks saga víðförla 76, 77, 86, 159
empire 103, 116, 119, 138, 156, 163
Byzantine 116
Holy Roman 50, 103, 163
Roman 50, 83, 84, 90–91, 93, 102–03, 116, 119, 126
see also translatio
emperors 73, 90
encyclopedias 15, 16, 17, 18, 23, 24, 32–33, 37–40, 42, 43, 44, 47, 51, 53, 58, 60–61, 63, 64, 67, 71, 73n87, 86, 88, 104, 114, 118, 129, 133, 134, 141, 143, 152, 157–59, 163, 164, 169n40, 171, 172
England 42, 94, 145, 156–57
see also Britain
Eridanus 145–47
Essex 100
Ethruria 146–47
Etymologiae libri XX see Isidore, *Etymologies*
etymology 48, 76n100, 95n47, 96n47, 99, 110n21, 119, 123n72, 149, 171
Eurocentrism 137
Europa, daughter of Agenor 95
Europe 13–14, 16–24, 25, 26, 30, 31, 32, 37, 45–47, 48, 49n41, 51, 62, 65–67, 69, 77–79, 84, 88–89, 90, 93, 95, 99, 102–05, 107–14, 116, 122–27, 129, 130, 134–38, 142–43, 149–52, 155–65, 170–72
as identity 19n21, 22, 165
as myth 19, 122
boundaries of 13–14, 28–30, 32, 45n30, 48, 93, 96, 106, 108, 115, 120–24, 126–27, 137, 163
marvels of 117–18, 120
European Union 22, 115n38
Europs (king) 95–96, 123n72, 171

Fazio degli Uberti 32, 53, 75, 79, 129, 136, 140–43, 159
Il Dittamondo 32, 53, 75, 79, 136, 143–43, 159
Feodosia 95–97
fire 37, 45, 49, 69n73, 74
see also four elements
Fóstbrœðra saga 76, 86, 159
four elements 37, 45–46, 73–74
fragments 23, 66, 76, 85, 158–59
France 41, 49n41, 57, 69, 75, 90, 91, 107, 114, 120, 136, 143, 151n66, 156n3
Kings of 49n41, 73, 74, 114
French 38, 43, 55, 57, 59, 64–67, 77, 89n25, 107–08, 109, 112, 118, 127, 138, 141n40, 150
see also Gossouin de Metz

Frutolf of Michselsberg 50n49, 95n47
fuzziness 17, 30, 32, 33, 108, 117, 122, 136, 155
fuzzy logic 18, 120
fuzzy sets 13, 17, 28, 30–32, 33, 44, 48, 61, 83, 90n30, 93, 104, 105–11, 113, 116–18, 121–27, 129, 136, 149–50, 152, 155–62, 165n29, 172

Gaius Plinius Secundus, *see* Pliny
Galicia 98
Galileo Galilei 27
Gallia 49n41, 123n67, 146–47, 148, 149
Belgica 148, 149
Ludgunensis 149
Narbonensis 149
Gascony 112
Gaul *see* Gallia
Gautier de Metz *see* Gossouin de Metz
Gautier de Châtillon *see* Walter of Châtillon
General Estoria 123, 133, 165n29
genres 16, 17, 23, 37–39, 47, 53, 61–62, 68, 74, 83, 85, 99, 132–34, 157–59, 161, 164
geocentric system 50
geography 13–18, 20–21, 23, 25, 26, 28–30, 32–33, 37, 39, 44, 45, 47, 48, 62, 64, 73, 74, 76, 77–78, 83, 85–87, 88, 91, 96, 99, 101, 103, 105, 106, 111–113, 115, 120, 126, 129–30, 134, 136, 137, 140–142, 149, 151, 155–57, 160–63, 165–67, 168, 170, 171
Classical 45, 92, 93, 96, 98
geographic object 14–15, 18, 21–23, 28, 33, 49, 62, 119, 122, 148, 156, 161, 164, 172
see also historico-geographical model
Gerald of Wales 119, 157n6
Germania Inferior 49n41, 85
Germania Superior 49n41, 92
Germany 41–42, 62, 65, 85, 89, 90–94, 143, 151n66, 173
Gervase of Tilbury 47n33, 77, 83, 99–103, 113, 114, 159–60, 163
Otia imperialia 47n33, 77, 99–101, 104
Glastonbury 44
Godfrey of Viterbo 78, 159
Gossouin de Metz 26n44, 53, 59, 60, 68, 69, 72, 74, 108, 112–16, 118, 120, 127, 132, 157, 159,
L'Image du monde 26n44, 32, 53, 55, 57, 59, 60, 68–74, 70n78, 72n84, 77, 84n3, 96n48, 103, 105–08, 111–22, 124n80, 127, 130, 132, 136, 138–39, 149–50, 156–60, 163, 167, 169
Gothia 89, 90, 110
Greece 49n41, 73, 108–09, 112, 116, 124n78, 143, 151n66
Greek (language) 141n40, 147
Greeks 109n13, 110n19, 115n38
Guillem Soler 20n25

Hadrian, emperor 90
Hagin Deulacres 71
Heaven 45, 50, 51
Hektaios of Milet 45n30
Hell 48
Hellespont 108, 173n8
Helpericus 88
Henri/Honoré 'Angloy' 40n11
Henry the Black of Bavaria 44, 91
Henry of Blois, abbot of Glastonbury, bishop of Winchester 44
Henry of Huntingdon 40, 41n11, 44, 64
Hercules, pillars of 13–14, 33, 83, 106, 127
Hesperides 125
Hircania 173n6
Historia Miscella 50n49
historico-geographical model 85, 106, 111, 115–19, 121, 127, 133, 131, 150, 160, 163–64
history 18, 26, 28, 47, 58, 69, 103, 115–16, 119, 126n85, 133, 137, 138, 142, 164, 165, 170
ecclesiastical 50, 100
in *Imago mundi* 50, 73n85, 100
of Africa 116, 126n85, 165n29
of Germany 93
of ideas of borders 29, 109, 127–28
of ideas of Europe 19–20, 21, 109, 119n38, 126n85, 164
of reading 132–33, 134
of science 27–28, 45
of Spain 123
periodisation of 18, 21n27, 27, 29, 167–68
self-other dichotomy in 118n50
synchronic presentation of 26, 45, 85, 150
universal 50, 99, 115–16, 133, 137
see also borders, chronotope, *Imago mundi*, space-time, time
hodoeporical descriptive technique 17, 33, 106, 109, 129, 132, 135–53, 159–60
Honorius Augustodunensis 15, 37, 38, 40–44, 50–51, 53, 60, 63–64, 67, 72, 73, 75, 78, 83–84, 86–94, 96, 98–104, 110, 125–26, 141–43, 149, 155–56, 158–61, 163, 168, 170–71
De luminaribus ecclesiae 41n11
De solis affectibus 72–73
Elucidarium 16n12, 43–44, 63, 72, 76, 86
Gemma animae 43n21, 46n33
Summa totius 50, 72–73
See also Autun; *Imago mundi*
Hermannus Contractus 50n49
Hyginus 88

Ibiza 48
Iceland 67, 86, 137
Ikaria 124
illustrations 25–26, 46n33, 59, 69–70, 111
see also cartography
Imago mundi 13, 15–18, 23, 25–26, 28, 30–33, 37–51, 53–57, 59–72, 74–79, 83, 85–101, 103–11, 113–14, 116, 119–21, 123–25, 127, 129–32, 134–45, 147–48, 150–72
authorship 37, 40–41, 44, 63, 158, 168
mis-attributions of 40–41, 44, 64, 158, 168

Book I 25, 37, 44, 45–46, 49–50, 51, 60, 62, 65–69, 73n87, 74, 78, 95, 105, 107–10, 125, 148, 155, 157
Book II 50
Book III 46n33, 50, 65n51
manuscripts 25–26, 40–42, 44, 56, 64, 69, 70, 77, 98n54, 135, 144–47, 162, 166–67, 171
Augsburg, Staats- und Stadtbibliothek 2 Cod. 108 44n24, 167, 169
Cambridge, Corpus Christi College MS 66 26n44, 98n54, 145–47
London, Lambeth Palace MS 371 146–47
Munchen, Bayerische Staatsbibliothek, clm. 536 49n44
structure 17, 45–51, 101, 106, 125, 135, 147–52, 159
translations/adaptations
Anglo-Norman 40n11, 64, 68, 77, 96–97
Castilian *see Semeiança del mundo*
Catalan *see* Ecumene Chart
French *see* Gossouin de Metz
German 65, 92–94, 144n51, 153, 159
Hebrew 70–71, 159, 169
Italian 56, 59, 62–63, 65, 70, 159, 163
see also Brunetto Latini
Old Norse 66, 76, 83, 85, 159
Welsh *see Delw y Byd*
versions of 42, 145–48
See also Honorius Augustodunensis
India 69n70, 76, 111n23, 130–31
incunabula 155, 166, 168, 170
see also print
Inn (river) 92
Iohannes Scotus *see* John Scott Eriugena
Ireland 41, 118, 124, 125, 136, 140,n35, 143, 153n6
Isidore of Seville 16n11, 23, 40–41, 45, 46–47n33, 49–51, 63–64, 75n99, 86, 88–89, 91, 98–99, 121–22n63, 123n73, 124, 137n25, 149, 157
Etymologies 41n11, 45, 46n33, 51, 63n36, 86, 89, 96n47, 98
Islam 113
islands 37, 40n9, 48, 78n109, 99, 106, 121, 123–25, 127, 129, 131, 136, 152, 162
Isocrates 45n30
Istria 151
Italian *see Imago mundi*
Italy 49n41, 65, 75, 90, 101–02, 112, 113n31, 129, 136, 143–52, 159, 163
itinerary 17, 32, 106, 129, 142–44, 147, 149–50, 153, 160
see also hodoeporical descriptive technique
Iudea 90

Jacob van Maerlant 58, 74–75, 136n22, 159, 164
Alexanders geesten 47n31, 58, 74, 136 n.22, 159
Jerusalem 102, 112
Juan de Mena 58, 63n35, 75, 136, 159, 167–68
Laberinto de Fortuna 47n31, 58, 63n35, 75, 136, 158–59, 168
John of Saint-Victor 78, 159

Kaffa 96
Karpathos 124
Kerch Straight 96–97

La lumiere as lais 43
La Petite Philosophie 40n11, 68, 159
Lambert of Saint-Omer 29n53
Lateran 76n100, 146
Latin (language) 15, 32, 39, 43, 44, 47, 53–56, 58, 61, 62, 64–67, 72, 77–78, 91, 93, 97–99, 101, 107–10, 114, 119, 122, 130–31, 133, 136, 138, 144–47, 150–52, 159, 162–64, 168
livre de clergie 57, 68, 107, 113, 157, 169
see also *clergie*
Latini, Brunetto *see* Brunetto Latini
Longobardia 145–46, 147, 149
Louis IX, king of France 69
Lusitania 98
Lybia 90, 110–11n21, 143n45

Macrobius 88
Maeotian Swamps 94–96, 142
mapa del mond see Belcalzèr, Vivaldo
mappae mundi 24–27, 29–30, 37, 40, 47–48, 59, 73, 74, 77–78, 85, 106, 111, 120, 124, 127, 138–40, 149–50, 167
see also cartography; illustration; world, tripartite structure of
Marco Polo 84
Martianus Capella 88
Mauretania Caesarensis 90
Mauretania Stifensis 90
Mauretania Tinguitana 90, 121
Mediterranean 17n15, 37, 46–48, 93–94, 105–07, 110, 120–27, 129, 152, 162, 165n29
Melos 124
Mena, *see* Juan de Mena
Meotides paludes see Maeotian Swamps
Meroë (island) 125
Michael Scot 78, 159
Minkowski, Hermann 127
mirror of princes genre 74, 99
Moesia 89
Murcia *see* Cartagena 98

Naxos 124
Newton, Isaac 27
Nicomedia 173n7
Nile 46–47n37, 106, 122
Noah, sons of 46–47n33, 73
Nomenclature 84, 87–93, 101, 119, 160
archaic 83, 93, 160
Roman 33, 83, 88–89, 98
see also provincia

Noricum *see* Bavaria
Norway 85, 86

oceans 45, 49, 107n5, 147
Old Norse 43, 66–67, 83
literature 66–67, 76, 85–87, 134–35, 137, 159
orality 131–32, 134
see also aurality
Orientalism 114–15, 117, 120
orientation 14, 23, 137–40, 142, 143, 147
Orosius 23, 40, 45n30, 50n49, 88–89, 113, 137n25
Ostia 100
Otto IV, emperor 77, 99
Otto of Freising 95n47
Ovid 173n8

Padus see Po (river)
Palestrina 100
Pamphylia 173n8
Pannonia 89
Paradise *see* Earthly Paradise
Paros 124
patronage 44, 64, 162, 165
Pentapolis 111n21
Perdita (island) 40n9, 125
Perot de Garbelei 58, 68, 159
Pierre d'Ailly 60n29, 78, 166, 171n49
Ymago mundi 60n29, 78, 166
Pillars of Hercules, *see* Hercules, pillars of
place-names *see* nomenclature
Pliny 88
Po (river) 145–49, 151
Poincaré, Henri 27
Polyhystor see Solinus
Porto 100
print 23, 40, 59, 69–71, 155, 166n30, 167–71
See also incunabula
provincia 83, 88–93, 101, 102, 110n21
Provinciale Romanum 101–02
Ptolemy 50, 142n42, 167
Geographia 167
Pseudo-Alcuin 88
Pseudo-Bede 88
see also Bede, the Venerable
Pyrenees 122

Rabanus Maurus 16n11, 88
Ranulf of Higden 29n53, 111, 113
Polychronicon 29n53, 111, 113
Ratispona see Regensburg
readers 17, 24, 32–33, 63, 70n77, 84n3, 94, 98, 112, 129–133, 137–40, 149, 156–57, 162, 171–72
Regensburg 41–42, 44, 91–93
regio 89–90, 92, 107n5, 109
relativity 27–28
Renaissance 27n47, 40n9, 98, 167, 171
Renus *see* Rhine
rewriting 54–58, 60, 62, 64, 67–68, 73, 74, 77, 84–85, 88, 95, 98–99, 101, 103, 106, 130, 134–38, 140, 150, 153, 157, 162–65, 167, 170, 172
see also adaptations; translation
Rhine 145–49, 151
Rhodes 124
Rhône 151
Rieti 100–01
Ripheian Mountains 95, 110
rivers 29n53, 45, 92, 95, 106, 110, 131, 147–48, 151
Robert d'Artois 69
Roman Empire *see* empire
Roman Forum 99
romance 55, 61, 134, 141, 158
Romans 110
Rome 47, 76n100, 94, 99–100, 116, 146–49, 151
see also empire, Roman
Romulus 76n100, 146
Rudolf of Ems 95n47, 163
Ruggiero 13–14, 18, 32–33, 37, 104, 106, 127, 129, 143

Sabina 100
Sallust 113
Salzburg 92
Samos 124
Sardinia 48, 125, 151
Scandinavia 67, 83, 85, 134
Semeiança del mundo 40n9, 41n11, 43, 53, 60, 63–64, 74, 77, 98, 121, 122, 123–24, 130–34, 136, 157–58, 159, 160, 163
sets *see* fuzzy sets
Sicily 48, 100, 124, 151
six ages of the world *see* world, six ages of
Solinus 73n85, 88, 142n42, 143n45
space-time 13, 24, 26–28, 45, 133, 156, 164
Spain 30, 49n41, 62, 64, 75, 90, 94, 98, 112, 113n31, 121–24, 129, 132, 134, 135, 136, 143, 149, 151n66, 157, 163
Speculum mundi see *Imago mundi*
Stjórn 86
Stoechades 124
Styria 92
Swabia 92–93
Syria 112

Tanais see Don (river)
Tanaus (king) 95
Tarragona 98
Tenedos 124
Thebaid 90, 103
Theodosia *see* Feodosia
Thomas of Cantimpré 39n6, 78, 159
time 13, 19n73, 22, 23, 24, 26–28, 32–33, 45, 48, 50, 54, 58–59, 83, 85, 87, 94, 99, 115–16, 126, 133, 137, 146, 156, 160, 164–65
see also chronotope; history; space-time
Tinguitania 98, 121

Tivoli 100
T-O maps *see mappae mundi*
toponyms *see* nomenclature
translatio 55, 57, 104, 116
  imperii 47, 48, 56, 73, 116, 119, 137–38, 156
  studii 53, 55, 57, 69, 116, 119, 138, 156
  studiorum 56
translation 13, 15, 16, 17, 23, 24, 30, 32, 33, 39, 42–43, 45, 53–68, 69n70, 70, 71, 73n87, 74–75, 76, 77–79, 83, 86, 94, 96–98, 107, 109, 118, 120, 123, 125, 127, 129, 132, 135n19, 136, 139, 144–45, 150, 153, 157–61, 163–64, 167, 168, 171
  concept of 24, 30, 53–55, 56, 57, 58, 60–61, 62, 67, 77, 104n80, 132, 157, 160
  translater 55
  see also adaptations; *Imago mundi*, translations/adaptations; rewriting
travel 13–14, 42, 86, 100, 115n38, 125, 129–30, 140, 142n42, 143, 147, 162–63
  armchair 32, 33, 83, 125, 127, 129–32, 150, 160, 162
  see also hodoeporical descriptive technique; itinerary
tripartite world *see* world, structure of
Trojan origin legend 47, 87
Troy 47, 146, 148
Tuscia 90, 145–46
Tusculum 100

Uberti, Fazio degli *see* Fazio degli Uberti
Umbria 147–48
Urban II, pope 20n25

Varro 113
Venetian (dialect) 65
Venice 146–49
Vienna Woods 92
Vincent of Beauvais 39n6, 57, 78, 159, 169n40, 171n49
Virgil 142n42
Vivaldo Belcalzèr *see* Belcalzèr, Vivaldo

Wales 26n44, 62, 65, 75, 116, 119, 132, 144
  see also Delw y Byd; Welsh
Walter of Châtillon 74
  Alexandreis 74
water 37, 45, 49, 74, 122, 128, 161
  see also four elements; rivers
Welsh 26n44, 42n16, 43, 64–65, 77, 96–97, 98n53, 109n16, 116, 132, 144–45
  see also Delw y Byd; Wales
west 14n3, 29, 44, 48, 58, 100, 109n13, 115–16, 118–20, 126n85, 138, 143, 151, 157
William of Malmesbury 20n25
world 14n3, 15, 23, 24, 25, 31, 33, 37–38, 44, 46n33, 49, 51, 67, 73, 74, 85, 87, 88, 93–94, 98–100, 113n32, 120–22, 126–27, 129, 131, 137, 143, 147, 150, 152n69, 153, 156–57, 160, 166–67, 170
  as egg 66
  cumulative description of 49, 56, 105, 106, 110
  six ages of 50, 73
  structure of 23, 25, 30,n58, 37, 45–48, 50, 76, 85, 87 nn. 14 and 15, 105–12, 115–17, 120, 122, 123–27, 130, 138–40, 142, 162
  sequence of parts of 25, 33, 37, 46–47, 51, 54, 70n77, 73, 101, 106, 108, 112, 125, 129, 130, 135, 137–40, 142–47, 151–52, 153, 156, 157, 162
  see also deviser; division; hodoeporical descriptive technique
worldview 23, 24, 28, 31, 47, 67n58, 87, 88, 119, 120, 156, 165, 166, 171

Ymagine del mondo 70, 159
Ystorya Bown o Hamtwn 132
  see also Welsh
zodiac 45

For Product Safety Concerns and Information please contact our EU representative GPSR@taylorandfrancis.com Taylor & Francis Verlag GmbH, Kaufingerstraße 24, 80331 München, Germany

Printed and bound by CPI Group (UK) Ltd, Croydon, CR0 4YY
14/09/2026
02216224-0013